Engineering Physics II

Dr.A. John Peter,

Associate Professor,

Department of Physics,

St.Anne's College of Engineering and Technology, Panruti.

Published by

Engineering Physics II

Copyright © 2017 by Bonfring

ISBN 978-93-86176-54-7

Author

Dr.A. John Peter

Bonfring

309, 2nd Floor, 5th Street Extension, Gandhipuram,

Coimbatore-641 012.

Tamilnadu, India.

E-mail: info@bonfring.org

Website: www.bonfring.org

Phone: 0422 4213231

Dedicated to

My Parents

Preface

Today's technological developments are the results of the joint efforts of physicists and engineers. A study of physics is, therefore, indispensible for students of engineering in order to excel in their respective fields. This book has been written with a view to provide a comprehensive text on Engineering Physics for the first year BE/B.Tech degree students studying in engineering college affiliated to the anna university steam. This book covers the topic like conducting materials, semiconducting materials, magnetic materials, super conducting materials, dielectric materials and advanced engineering materials such as metallic glasses, shape memory alloys, nano materials and carbon nano tubes. Conducting material concepts have been explained in detail in **Chapter 1**. Various types of semiconducting materials along with their applications are covered in **Chapter 2**. **Chapter 3** describes the basics of various types of magnetic and super conducting materials and their applications. **Chapter 4** discusses dielectric materials and their applications. **Chapter 5** elucidates advanced engineering materials such as metallic glasses, shape memory alloys, nano materials and carbon nano tubes along with their applications. The text presents the fundamental principles of physics and their applications in a simple language. A lucid writing style has been followed, which makes easier for students to understand the concepts. Equal attention is devoted to elucidating engineering applications of the various topics in every chapter. To summarize, the salient features of this book are

- Complete syllabus coverage.
- Simple and lucid writing style.
- Rich pedagogy.

Dr.A. John Peter

Acknowledgement

I profusely thank **Rev. Mother Victoria, SAT.,** Secretary, **Dr.R. Arokiadass,** Principal, **Rev. Sr. Gnana Jency Salete Mary, SAT.,** Vice Principal and **Rev. Sr. A. Avila Therese, SAT.,** Head, Department of Science and Humanities of St. Anne's College of Engineering and Technology for their permission and constant encouragement in bringing out edition of this book.

The academic nurturing bestowed by **Dr. I. B. Shameem Banu,** Professor of Physics, B. S. Abdur Rahman University, Vandalur, Chennai is highly praiseworthy and it was her expertise and knowledge that led me to move boldly towards this goal. I take this opportunity to extend my sincere gratitude to her for her patience and constructive support in making this book.

My sincere thanks to **Dr. D. Arivouli,** Professor, Crystal Growth Centre, Anna University, Chennai and **Dr. D. Easwaramoorthy,** Professor and Head, Department of Chemistry, B. S. Abdur Rahman University, Vandalur, Chennai, to **Dr. I. Raja Mohamed** professor and Head, Department of Physics, B. S. Abdur Rahman University, Vandalur, Chennai, **Dr. M. S. Ramachandra Rao,** Professor, Materials Science Research Centre and Department of Physics, Indian Institute of Technology (IIT) Madras, Chennai and **Dr. P. Prabhakar Rao,** Scientist, Materials and Minerals Division, National Institute for Interdisciplinary Science and technology (CSIR), Thiruvananthapuram who played pivotal role in the initiation and progress of this book.

I would like to record my love and affection to my parents, wife and sweet daughter who are my strength in all my endeavours.

I feel highly blessed by the Almighty in making my dream come true.

UNIT-I

CONDUCTING MATERIALS

Abstract

After studying this unit, one can acquire:

1) Understanding about the conduction properties of metals.

2) Knowledge on various theories of electrons in solids.

3) Knowledge on the different factors, affecting the conductivity of metals and the energy of electrons in metals.

4) Proper knowledge on the distribution and concentration and emission of electrons.

Understanding about the variation of Fermi level with temperature and concentration of impurities in semiconductors.

1.1.　Introduction

Conducting materials have high electrical and thermal conductivity. These materials have very low resistivity which increases with temperature. Metals like gold, silver, copper and aluminium are good conductors of electricity. Conducting materials are mainly used for electrical power transmission.

In this chapter we will discuss the classical and quantum free electron theories to explain the conductivity of these materials.

1.1.1.　Terms and Definitions

1) *Conductors*

Experimental measurements showed that the metals and their alloys exhibit large electrical conductivity in the order of $10^8 \, \Omega^{-1} \, m^{-1}$.

Hence they are known as conductors. Conducting materials are the materials having high electrical and thermal conductivities.

Low resistive materials are also generally known as conducting materials.

2) *Bound Electrons*

All the valence electrons in an isolated atom are bound to their parent nuclei. They are called as 'bound electrons'

3) *Free Electrons*

In a solid, due to the boundaries of neighbouring atoms overlap each other, the valence electrons find continuity from atom to atom.

Therefore, they can move easily throughout the solid. All such valence electrons of its constituent atoms in a solid are called free electrons.

4) *Difference between Ordinary Gas and Free Electron Gas*

The molecules of ordinary gas are neutral. But, the free electron gas is charged. The density of molecules is smaller than the density of free electrons.

5) *Electric Field (E)*

The electric field (E) of a conductor having uniform cross section is defined as the potential drop (V) per unit length (l).

$$E = V/l \quad \text{Unit - Vm}^{-1}$$

6) *Current Density (J)*

Current density (J) is defined as the current per unit area of cross section of an imaginary plane hold normal to the direction of flow of current in a current carrying conductor.

If 'I' is the current, and 'A' is the area of cross section, then current density is given by,

$$J = I/A \quad \text{Unit - A m}^{-2}$$

1.2.   Conducting Materials

Conducting materials are classified in to three major categories based on the conductivity.

1)   Zero resistive materials.
2)   Low resistive materials.
3)   High resistive materials.

1) *Zero Resistive Materials*

The super conductors like alloys of aluminium, zinc, gallium, niobium, etc., are a special class of materials. These materials conduct electricity almost with zero resistance below transition temperature. Thus, they are called zero resistive materials.

These materials are used for saving energy in the power systems, super conducting magnets, memory storage elements etc.

2) Low Resistive Materials

The metals like silver, aluminium and alloys have high electrical conductivity. These materials are called low resistive materials.

They are used as conductors, electrical conduct etc., in electrical devices and electrical power transmission and distribution, winding wires in motors and transformers.

3) High Resistive Materials

The materials like tungsten, platinum, nichrome etc., have high resistive and low temperature co-efficient of resistance. These materials are called high resistive materials. Such a metals and alloys are used in the manufacturing of resistors, heating elements, resistance thermometers.

The conducting properties of solid do not depend on the total number of the electrons available because only the valance electrons of the atoms take part in the conduction. When these valance electrons detached from the orbit they are called free electrons or conduction electrons. In a metal, the number of free electrons available is proportional to its electrical conductivity. Hence, electronic structure of a metal determines its electrical conductivity.

1.3. Electron Theory of Solids

We know that the electrons in the outermost orbit of the atom determine the electrical properties in the solid. The free electron theory of solids explains the structure and properties of solids through their electronic structure. This theory is applicable to all solids, both metals and non metals. It explains:

1) The behavior of conductors, semiconductors, and insulators.
2) The electrical, thermal and magnetic properties of solids. So far three electron theories have been proposed.

1) Classical Free Electron Theory

It is a macroscopic theory, proposed by Drude and Lorentz in 1900. According to this theory, the free electrons are mainly responsible for electrical conduction in metals. This theory obeys laws of classical mechanics.

2) Quantum Free Electron Theory

It is a microscopic theory, proposed by sommerfeld in 1928. According to this theory, the electrons in a metals move in a constant potential. This theory obeys laws of quantum mechanics.

3) Zone Theory or Band Theory of Solids

Bloch proposed this theory in the year 1928. According to this theory, the free electrons move in a periodic potential. This theory explains electrical conductivity based on the energy bands.

1.4. Classical Free Electron Theory

Free electron theory of metals was proposed by P. Drude in the year 1900 to explain electrical conduction in metal. This theory was further extended by H.A. Lorentz in the year 1909.

1.4.1. Drude & Lorentz Theory

1) Principle

According to this theory, a metal consists of a very large number of free electrons. These free electrons can move freely throughout the volume of the metal. They are fully responsible for the electrical conduction in the metal.

2) Explanation

We know that an atom consists of a central nucleus with positive charge surrounded by the electrons of negative charge. The electros in the inner shells are called core electros and those in the outermost shell are called valence electrons as shown in figure. 1.1 .

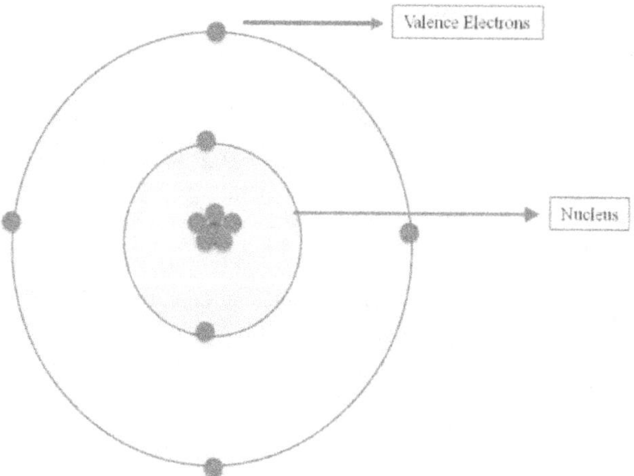

Figure 1.1: Schematic Diagram of an Atom

1.4.2. *Postulates of Free Electron Theory*

Drude assumed that the free electrons in a metal form an electron gas. These free electrons move randomly in all possible directions just like the gas molecules in a container.

In the Absence of Electrical Field

1) When an electrical field is not applied, the free electrons move everywhere in a random manner. They collide with other free electrons and positive ion core as shown in figure 1.2. This collision is known as elastic collision.

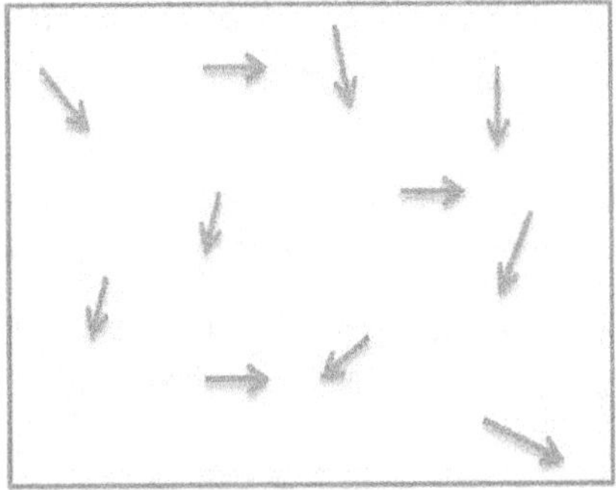

Figure 1.2: Movement of Electrons in the Absence of Electric Field

In the Presence of Electric Field

1) When the electrical field is applied, the electrons get some amount of energy from the applied electric field and they begin to move towards the positive potential as shown in figure 1.3. (In the opposite direction to the applied electric field).

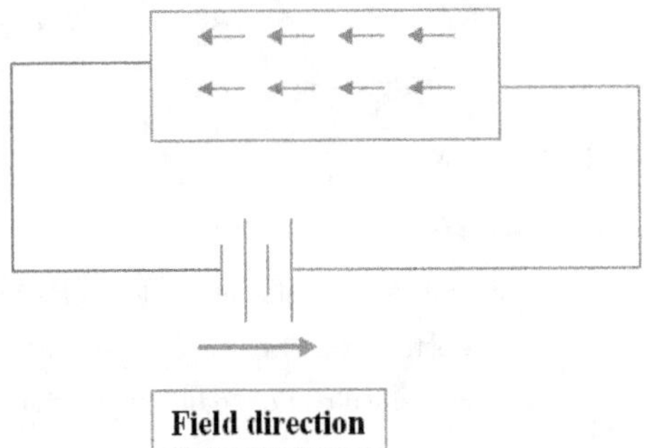

Figure 1.3: Movement of Electrons in the Presence of Electric Field

2) Since electrons are assumed to be a perfect gas, they obey the laws of kinetic theory of gases.

3) Drift velocity (v_d): It is defined as the average velocity acquired by the free electrons in a metal in a particular direction by the application of an electrical field.

4) Mean free path (λ) : The average distance travelled by a free electron between any two successive collisions in the presence of an applied electric field is known as mean free path. It is the product of drift velocity of free electron and collision time (τ_c).

$$\lambda = (v_d) \times (\tau c)$$

5) Collision time (τc) : The average time taken by a free electron between any two successive collisions is known as collision time.

$$\tau_c = \lambda / v_d$$

6) Relaxation time (τ) : The average time taken by a free electron to reach its equilibrium position from its disturbed position due to the application of an external electrical field is called relaxation time. It is approximately equal to 10^{-14} second.

1.4.3. Electrical Conductivity

Definition

The amount of electrical charge conducted (Q) per unit time across unit area (A) of a solid per unit applied electric field (E) is called the electrical conductivity. It is denoted by σ. It is given by

$$\sigma = Q / tAE$$
$$\text{If, } t = 1 \text{second } E = 1 \text{ volt, } A = 1 \text{ metre}^2$$
$$\sigma = Q$$
$$\sigma = Q / tAE = J / E$$

Where J is the current density and it is given by Q / tA.

Also, $J = \sigma E$ (according to ohm's law)

Expression for Electrical Conductivity

We know in the absence of external electric field, the motion of electrons in a metal moves randomly in all directions. When electric field (E) is applied to electron of charge 'e' of a metallic rod, the electron moves in opposite direction to the applied field with velocity v_d. This velocity is known as drift velocity is shown in figure 1.4.

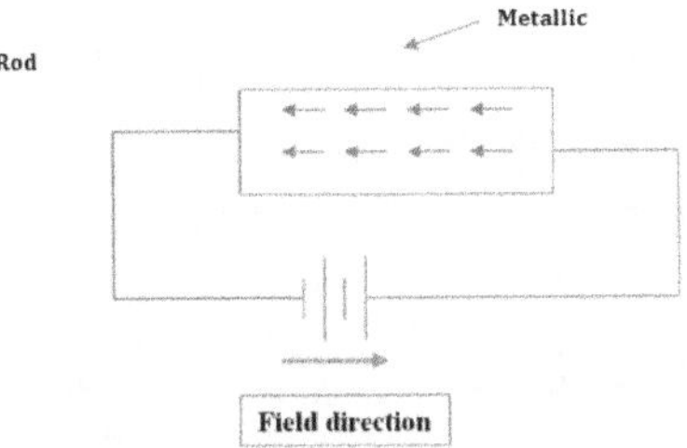

Figure 1.4: Movement of Electrons in the Presence of Electric Field

Lorentz force acting on the electron $F = eE$ \quad (1)

This force is known as driving of the electron. Due to this force, the electron gains acceleration 'a'. From Newton's second law of motion, the force

From equations 1 & 2 \quad $F = ma$

$$ma = eE \ (or) \quad (2)$$
$$a = eE / m \quad (3)$$

The acceleration of electron is given by

Acceleration (a) = Drift velocity (v d) / Relaxation time (τ)

$$a = v_d / \tau$$
$$(or) \ v_d = a\,\tau \quad (4)$$

Substituting equation (3) in (4)

$$v_d = [\,e\,\tau / m\,]\, E \quad (5)$$

Where 'σ' is the electrical conductivity of the electron. But, the current density in terms of drift velocity is given as

$$J = ne\, v_d \quad (6)$$

Substituting equation (5) in (7), we have

$$J = ne\, [e\,\tau / m]\, E \quad (7)$$

$J = [ne^2\,\tau / m]\, E$, in terms of effective mass m^* of an electron,

$$J = [ne^2\,\tau / m^*]\, E \quad (8)$$

From microscopic form of Ohm's law, the current density 'J' is expressed as,

$$J = \sigma\, E \quad (9)$$

On comparing equations (8) & (9), we have

Electrical conductivity

$$\sigma = ne^2 \tau / m^* \qquad (10)$$

From equation (10), we know that with increase of electron concentration 'n', the conductivity 'σ' increases. As m^* increases, the motion of electron becomes slow and the electrical conductivity.

1.4.4. Thermal Conductivity

Definition

It is defined as the amount of heat flowing per unit time through the material having unit area of cross section per unit temperature gradient.

$$(ie)., Q = K \, dT/ \, dx$$

Thermal conductivity of the material, $\quad K = Q/ \, dT/dx$

Q – Amount of heat flowing per unit time through unit cross sectional area,

dT/dx – Temperature gradient.

Expression for Thermal Conductivity

Let us consider a uniform rod AB with temperatures T_1 (Hot) at end A and T_2 (cold) at end B. Heat flows from hot end A to the cold end B. Let us consider cross sectional area C which is at a distance equal to the mean free path (λ) of the electron between the ends A and B of the rod as shown in fig.

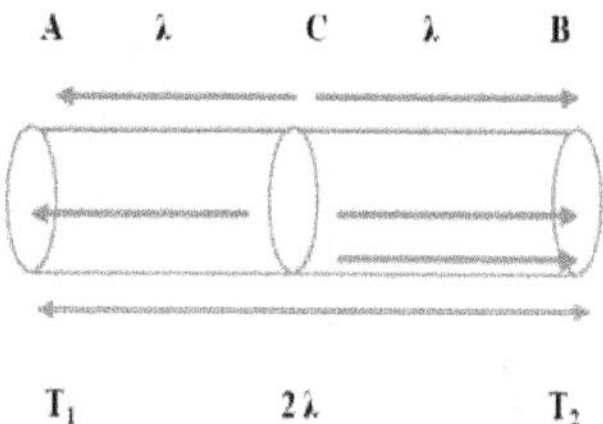

Figure 1.5: Schematic Diagram of Heat Flow in a Rod

The conduction electron per unit volume is n and average velocity of these electrons is v.

During the movement of electrons in the rod, the collision takes place. Hence, the electrons near A lose their kinetic energy while electrons near B gain kinetic energy.

At A, average kinetic energy of en electron

$$= 3/2 \, kT \qquad (1)$$

8

Where k- Boltzmann's constant and T is temperature at A.

At B average kinetic energy of the electron$=3/2\ k(T-dT)$ (2)

The excess of kinetic energy carried by the electron from A to B

$$= 3/2\ kT - 3/2\ k\ (T - dT)$$
$$= 3/2\ kT - 3/2\ kT + 3/2\ k\ dT \qquad (3)$$

Number of electrons crossing per unit area per unit time from A to B

$$= 1/6\ nv$$

The excess of energy carried from A to B per unit area in unit time

$$= 1/6\ nv \times 3/2\ kdT$$
$$= 1/4\ n\ v\ k\ dT \qquad (4)$$

Similarly, the deficient of energy carried from B to A per unit area per unit time

$$= -1/4\ n\ v\ kdT \qquad (5)$$

Hence, the net amount of energy transferred from A to B per unit area per unit time

$$Q = 1/4\ nvkdT - (-1/4\ nvkdT\ )$$
$$Q = 1/4\ nvkdT + 1/4\ nvkdT$$
$$Q = 1/2\ nvkdT \qquad (6)$$

But from the basic definition of thermal conductivity, the amount of heat conducted per unit area per unit time

$$Q = K\ dT/\lambda$$
$$[Q = K\ dT/\ dx;\ \lambda = dx]$$
$$1/2\ n\ v\ k\ dT = K\ dT/\lambda$$
$$K = \tfrac{1}{2}\ nv\ k\ \lambda \qquad (7)$$

We know that for the metals,

Relaxation time = Collision time

$$\tau = \tau_c = \lambda/v$$
$$\tau v\ = \lambda \qquad (8)$$

Substituting equation 8 in the equation 7, we have

$$K = \tfrac{1}{2}\ n\ v\ k\ \tau\ v$$
$$K = \tfrac{1}{2}\ n\ v^2\ k\ \tau \qquad (9)$$

Equation 9 is the classical expression for the thermal conductivity of metal.

1.4.5. *Widemann–Franz Law*

Statement

It states that for the metals, the ratio of thermal conductivity to electrical conductivity is directly proportional to the absolute temperature. This ratio is a constant for all metals at given temperature.

$$K / \sigma \alpha T$$
$$\text{Or } K / \sigma = LT$$

Where L is proportionality constant. It is known as Lorentz number. Its value is 2.44×10^{-8} $W\Omega K^{-2}$ at T = 293 K.

Derivation

Widemann – Frantz law is derived from the expressions of thermal conductivity and electrical conductivity of metal.

We know that,

Electrical conductivity of a metal $\sigma = ne^2 \tau / m \qquad (1)$

Thermal conductivity of a metal $K = \tfrac{1}{2} n v^2 k \tau \qquad (2)$

Thermal conductivity / Electrical conductivity

$$= \tfrac{1}{2} n v^2 k \tau m / ne^2 \tau \qquad (3)$$

We know that the kinetic energy of an electron

$$\text{Or } \tfrac{1}{2} mv^2 = 3/2 \, kT \qquad (4)$$

Substituting equation 4 in equation 3, we have

$$K/\sigma = 3/2 \, kT \cdot k / e^2$$
$$= 3/2 \, [k/e]^2 \, T$$
$$K/\sigma = LT \qquad (5)$$

where $L = 3/2 \, [k/e]^2$ is a constant and it is known as Lorentz number

$$K/\sigma \; \alpha \; T \qquad (6)$$

Thus it is proved that, the ratio of thermal conductivity to electrical conductivity of a metal is directly proportional to the absolute temperature of the metal.

Wildman – Franz law clearly shows that if a metal has high thermal conductivity, it will also have high electrical conductivity.

1.4.6. *Lorentz Number*

The ratio of thermal conductivity (K) of a metal to the product of electrical conductivity (σ) of the metal and absolute temperature (T) of the metal is a constant. It is known as Lorentz number and it is given by

$$L = K / \sigma$$

1.4.7. *Merits and Demerits of Classical Free Electrton Theory*

1) It is used to verify Ohm's Law.
2) The electrical and Thermal conductivities of metals can be explained by this theory.
3) It is used to derive Wiedemann- Franz law.
4) It is used to explain the optical properties of metals.

Demerits

It is a macroscopic theory

1) Classical theory states that all free electrons will absorb energy, but quantum theory states that only few electrons will absorb energy.
2) This theory cannot explain the Compton effect, photoelectric effect, paramagnetism, ferromagnetism etc.
3) The theoretical and experimental values of specific heat and electronic specific heat are not matched.
4) By classical theory K / σ T = constant for all temperatures, but by quantum theory K / σ T $\neq$ constant for all temperatures.
5) The Lorentz number by classical theory does not have good agreement with experimental value and it is rectified by quantum theory.

1.5. Quantum Theory

The drawbacks of classical theory can be rectified using quantum theory. In classical theory the properties of metals such as electrical and thermal conductivities are well explained on the assumption that the electrons in the metal freely moves like the particles of gas and hence called free electron gas.

According to classical theory, the particles of gas (electrons) at zero Kelvin will have zero kinetic energy, and hence all the particles are at rest. But according to quantum theory when all particles at rest, all of them should be filled only in the ground state energy level, which is impossible and is controversial to the pauli's exclusion principle.

Thus in order to fill the electrons in a given energy level, we should know the following.

1) Energy distribution of electrons
2) Number of available energy states
3) Number of filled energy states
4) Probability of filling an electron in a given energy state, etc.,

1.6. Fermi Dirac Distribution Function

The classical and quantum free electron theories failed to explain many electrical and thermal properties of solids. However, these properties can be easily understood using Fermi–Dirac statistics. Fermi–Dirac statistics deals with the particles having half integral spin. The particles like electrons are the examples of half integral spin and hence they are known as Fermi particles or Fermions.

Definition

The expression which gives the distributions of electrons among the various energy levels as a function of temperature is known as Fermi distribution function. It is the probability function F(E) of an electron occupying given energy level at absolute temperature. It is given by

$$F(E) = \frac{1}{1+e^{(E-E_F)/kT}} \qquad (1)$$

Where
E – Energy of the level whose occupancy is being considered
EF – Energy of the Fermi level
K –Boltzmann's constant
T – Absolute temperature

The probability value of F(E) lies between 0 and 1. If F(E) = 1, the energy level is occupied by an electron. If F(E) = 0, the energy level is vacant. If F(E) = ½ or 0 .5 then there is a 50% chance for the electron occupying in that energy level.

Effect of Temperature on Fermi Function

The effect of temperature on Fermi function F(E) can be discussed as following.

1) At 0 kelvin

At 0 kelvin, the electron can be filled only upto a maximum energy level called Fermi energy level (EF0), above EF0 all the energy levels will be empty. It can be proved from the following conditions.

1) When E<E_F, equation (1) becomes

$$F(E) = \frac{1}{1+e^{-\infty}} = 1/1 = 1$$

(i. e. 100% chance for the electron to be filled within the Fermi energy level)

2) When E>EF, equation 1 becomes

$$F(E) = \frac{1}{1+e^{\infty}} = 1/\infty = 0$$

(i.e zero% chance for the electron not to be filled within the Fermi energy level)

3) When E = EF, equation 1 becomes

$$F(E) = \frac{1}{1+1} = ½ = 0.5$$

(i.e 50% chance for the electron to be filled within the Fermi energy level)

This clearly shows that at 0 kelvin all the energy states below E_{FO} are filled and all those above it are empty. The Fermi function at 0 kelvin can also be represented graphically as shown in figure 1.6.

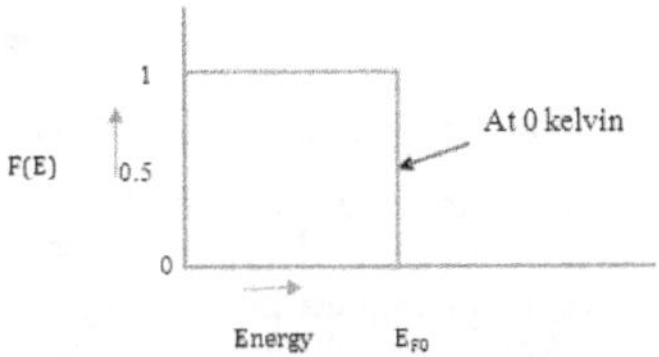

Figure 1.6: Representation of Fermi Function with Energy at 0 Kelvin

Fermi Energy and Its Importance Fermi Energy Level

Fermi energy level is the maximum energy level upto which the electron can be filled at ok.

Importance

1) Thus it act as a reference level which separates the vacant and filled energy states at 0k. It gives the information about the filled electron states and the empty states.
2) At 0k, blow Fermi energy level electrons are filled and above Fermi energy level it will be empty.

When the temperature is increased, few electrons gains thermal energy and it goes to higher energy levels.

Conclusions

1) In the quantum free electron theory, though the energy levels are discrete, the spacing between consecutive energy level is very less and thus the distribution of energy levels

seems to be continuous.

2) The number of energy levels N(E) that are filled with electrons per unit energy increases parabolic ally with increase of energy E as shown in figure 1.7.

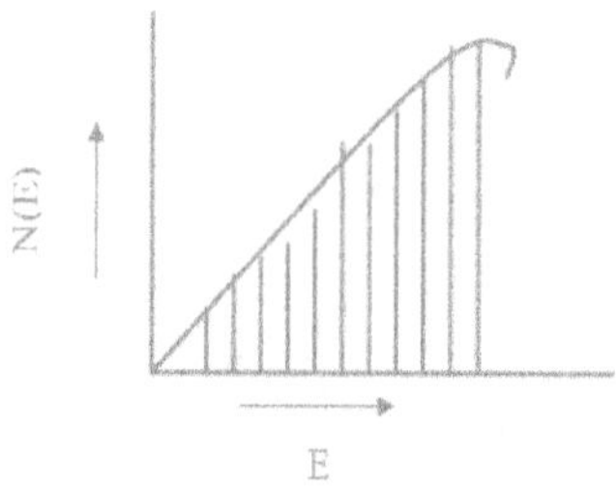

Figure 1.7: The Representation of Change in Energy of Electron at Room Temperature

1) Each energy level can provide only two states, namely, one for spin up and other for spin down and hence only two electrons can be occupied in a given energy states i.e., pauli's exclusion principle.

2) At T = 0, if there are N number of atoms, then we have N/2 number of filled energy levels and other higher energy levels will be completely empty.

3) This $(N/2)^{th}$ energy level is the highest filled energy level is known as Fermi energy level (E_{FO}).

4) The electrons are filled in a given energy level, according to pauli's exclusion principle (i.e) no two electrons can have the same set of four quantum numbers.

5) At room temperature, the electrons within the range of K_BT below the Fermi energy level will absorb thermal energy = K_BT and goes to higher energy states with energy E_{F0} +K_BT.

1.7. Density of States

The Fermi function F(E) gives only the probability of filling up of electrons on a given energy state, it does not give the information about the number of electrons that can be filled in a given energy state. To know that we should know the number of available energy states so called density of states.

Definition

Density of states Z(E)dE is defined as the number of available electron states per unit volume in an energy interval (dE).

Explanation

In order to fill the electrons in an energy state we have to first find the number of available energy states within a given energy interval. We know that a number of available energy levels can be obtained for various combinations of quantum numbers n_x, n_y, and n_z . (i.e) $n^2 = n^2_x + n^2_y + n^2_z$

Therefore, let us construct three dimensional space of points which represents the quantum numbers n_x, n_y, and n_z as shown in fig. in this space each point represents an energy level.

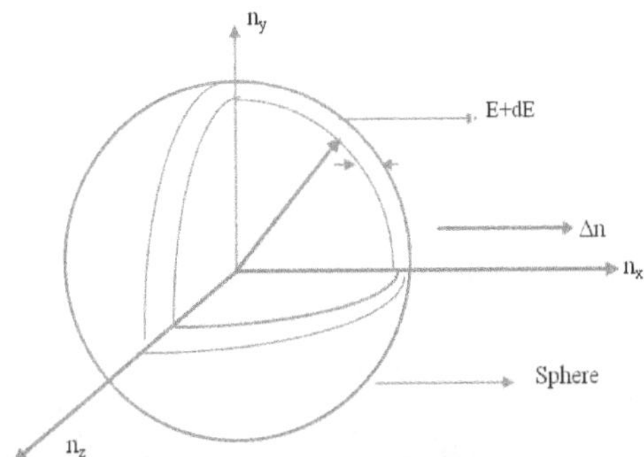

Figure 1.8: The Representation of Density of Energy States

1.7.1. Derivation of Density of Energy States

To find the number of energy levels in a cubical metal piece and to find number of electrons that can be filled in a given energy level, let us construct a sphere of radius 'n' in the space.

The sphere is further divided in to many shells and each of this shell represents a particular combination of quantum numbers (n_x, n_y, and n_z) and therefore represents a particular energy value.

Let us consider two energy values E and E + dE. The number of energy states between E and E + dE can be found by finding the number of energy states between the shells of the radius n and n + Δn, from the origin. The number of energy states within the sphere of radius n = 4/3 π n^3

Since nx, ny, and nz will have only positive values, we have to take only one octant of the sphere (i.e) 1/8 th of the sphere volume. The number of available energy states within the sphere of radius n = 1/8 [4/3 π n^3]

Similarly the number of available energy states within the sphere of radius n + dn

$$n + dn = 1/8\ [4/3\ \pi\ (n+dn)^3]$$

The number of available energy states between the shells of radius n and n + dn (or) between the energy levels

$$E \text{ and } E + dE = 1/8\ [4/3\ \pi\ (n+dn)^3 - 4/3\ \pi\ n^3\ ]$$

(i.e) The number of available energy states between the energy interval,

$$Z(E)dE = 1/8 \times 4/3\ \pi\ [n^3+dn^3+3n^2dn+3dn^2n-\ n^3]$$

Since the higher powers of dn is very small, dn^2 and dn^3 terms can be neglected.

$$Z(E)dE = \pi\ /\ 6\ (3n^2dn)$$

$$Z(E)dE = \pi\ /\ 2(n^2\ dn) \qquad\qquad (1)$$

We know the energy of the electron in a cubical metal piece of sides,

$$\ell = E = -\ \frac{n^2 h^2}{8ml^2}$$

Differentiating equation (2) we get,

$$n^2 = -\ \frac{8ml^2}{h^2} \qquad\qquad (2)$$

$$n = \left(\frac{8ml^2}{h^2}\right)^{1/2} \qquad\qquad (3)$$

$$2ndn = \frac{8ml^2}{h^2}\ dE \qquad\qquad (4)$$

$$ndn = \frac{8ml^2}{2h^2}\ dE$$

Equation 1 can be written as,

$$Z(E)dE = \pi\ /\ 2(n^2\ dn)$$

$$Z(E)dE = \pi\ /\ 2\ n\ (n\ dn)$$

Substituting equation (3) and (4) in the above equation we have,

$$Z(E)dE = \pi\ /\ 2\ \left(\frac{8ml^2 E}{h^2}\right)^{1/2}\ \left(\frac{8ml^2}{2h^2}\ dE\right)$$

$$Z(E)dE = \pi\ /\ 2 \times \tfrac{1}{2}\ \left(\frac{8ml^2}{h^2}\right)^{1/2}\ \left(\frac{8ml^2}{h^2}\right)\ E^{1/2}\ dE$$

$$Z(E)dE = \pi\ /\ 4\ \left(\frac{8m}{h^2}\right)^{3/2}\cdot E^{1/2}\ d\ E$$

$$Z(E)dE = \pi\ /\ 4\ \left(\frac{8m}{h^2}\right)^{3/2}\ l^3\cdot E^{1/2}\ d\ E$$

Here ℓ^3 represents the volume of the metal piece. If $\ell^3 = 1$, then we can write that the number of available energy states per unit volume (i.e) Density of states

$$Z(E)dE = \pi/4 \left(\frac{8m}{h^2}\right)^{3/2} \cdot E^{1/2}\, d \quad (5)$$

Since each energy level provides 2 electron states one with spin up and another spin down, we have. Density of states

$$Z(E)dE = 2 \times \pi/4 \left(\frac{8m}{h^2}\right)^{3/2} \cdot E^{1/2}\, dE$$

$$Z(E)dE = \pi/2 \left(\frac{8m}{h^2}\right)^{3/2} \cdot E^{1/2}\, dE \quad (6)$$

1.7.2. Calculation of Carrier Concentration

Let $N(E)\,dE$ represents the number of filled energy states between the interval of energy dE. Normally all the energy states will not be filled. The probability of filling of electrons in a given energy state is given by Fermi function $F(E)$.

$$N(E) = Z(E)dE \cdot F(E) \quad\quad\quad (7)$$

Substituting equation (6) in equation (7), we get Number of filled energy states per unit volume

$$N(E) = \pi/2\,[8m/h^2]\,E^{1/2}\,dE.\,F(E) \quad\quad (8)$$

$N(E)$ is known as carrier distribution function (or) carrier concentration in metals.

1.7.3. Calculation of Fermi Energy

Case (i) At T= 0K,

$$E_F = [3Nh^3/\pi(8m)^{3/2}]^{2/3} = h^2/8m[3N/\pi]^{2/3}$$

Case(ii) At T > 0K,

$$E_F = E_{F0}\,[1 - \pi^2/12(kT/E_{F0})^2]$$

1.7.4. Average Energy of an Electron at 0K

$$\text{Average energy} = \frac{\text{Total Energy}}{\text{Carrier Concentration}}$$

$$\text{Total Energy} = \pi/5h^3(8m)^{3/2}\,E_{F0}^{5/2}$$

$$\text{Carrier Concentration} = \pi/3h^3(8m)^{3/2}\,E^{3/2}$$

$$\text{Average energy} = \frac{\pi/5h^3(8m)^{3/2}\,E_{F0}^{5/2}}{\pi/3h^3(8m)^{3/2}\,E^{3/2}}$$

$$E_{avg} = 3/5\,E_{F0}$$

UNIT-II

SEMICONDUCTING MATERIALS

Abstract

After studying this unit, one can acquire

1) Understand about the nature and properties of semiconductors.
2) Acquire knowledge on carrier concentration and its variation with temperature in different types of semiconductors.
3) Acquire better understanding about Hall effect and its applications.

2.1.   Introduction

A semiconductor is a solid which has the energy band similar to that of an insulator. It acts as an insulator at absolute zero and as a conductor at high temperatures and in the presence of impurities. Semiconductor devices revolutionised the field of electronics and replaced the vacuum tube devices which were used earlier. The vacuum tube devices required external heating and occupied more space whereas the semiconductor devices operate at low voltage, have low power consumption and occupy less Space. In vacuum tube devices, the electrons were made to travel in vacuum between the cathode and the anode whereas in semiconductor devices, the electrons move within the solid only. Hence, the semiconductor devices are also known as solid state devices.

2.1.1.   General Properties of Semiconductors

1) The resistivity of a semiconductor is lesser then an insulator but more than that of a conductor.
2) It is in the order of 10^{-4} to 0.5 ohm metre.
3) Based on energy band, A semiconductor has nearly empty conduction band and almost filled valance band with very small energy gap ($\approx$ 1eV).
4) They are formed by covalent bonds.
5) They have an empty conduction band at 0K They have almost filled valance band.
6) They have small energy gap They posses crystalline structure.
7) They have negative temperature co efficient of resistance.
8) If the impurities are added to a semiconductor, its electrical conductivity increases. Similarly, if the temperature of the semiconductor increased, its electrical conductivity increases.

2.1.2. Elemental and Compound Semiconductors

The semiconductors are classified mainly into two types based on composition of materials.

1) Elemental semiconductors and
2) Compound semiconductors

1) Elemental Semiconductors

These semiconductors are made from a single element of fourth group elements of the periodic table. They are also known as indirect band gap semiconductors Example– Germanium, silicon.

2) Compound Semiconductors

Semiconductors which are formed by combining third and fifth group elements or second and sixth group elements in the periodic table are known as compound semiconductors. These compound semiconductors are also known as direct band gap semiconductors.

Example–1: Combination of third and fifth group elements

Gallium phosphide (GaP), Gallium arsenide (GaAs), Indium phosphide (InP), Indium arsenide (InAs).

Example–2: Combination of second and sixth group elements

Magnesium oxide (MgO), Magnesium silicon (MgSi), Zinc oxide (ZnO), Zinc sulphide (ZnS).

2.2. Types of Semiconductors

Based on the purity semiconductors are classified in to the following two types.

1) Intrinsic semiconductors
2) Extrinsic semiconductors

1) Intrinsic Semiconductors

A semiconductor in extremely pure form, without the addition of impurities is known as intrinsic semiconductors. Its electrical conductivity can be changed due to thermal excitation.At 0K the valance band is completely filled and the conduction band is empty. The carrier concentration (i.e) electron density (or) hole density increases exponentially with increase in temperature.

2) Extrinsic Semiconductors

A semiconductor in extremely impure form, with the addition of impurities is known as extrinsic semiconductors.

2.2.1. *Intrinsic Semiconductors–Electrons and Holes*

We know that, at 0K intrinsic pure semiconductor behaves as insulator. But as temperature increases some electron move from valance band to conduction band as shown in fig. therefore both electrons in conduction band and holes in valance band will contribute to electrical conductivity. Therefore the carrier concentration (or) density of electrons (n_e) and holes (n_h) has to be calculated.

1) *Carrier Concentration of Electrons in the Conduction Band of an Intrinsic Semiconductor*

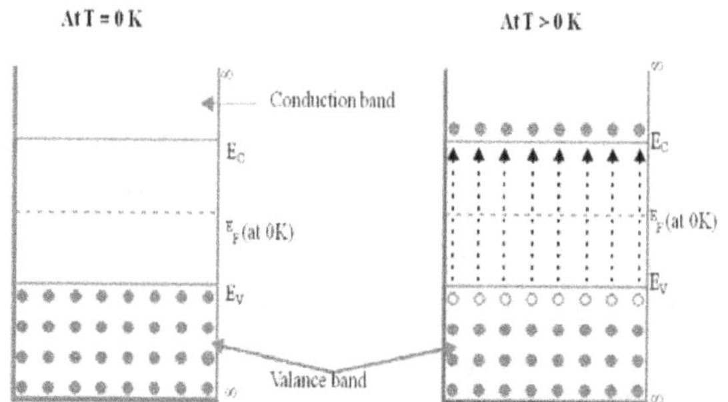

Figure 2.1: Energy Band Diagram of Intrinsic Semiconductor

Derivation:

The No of electrons in the energy level E & dE (dn) = Z(E) dE $\hspace{2cm}$ (1)

The No of electrons in the conduction band (n) = $\int_{E_C}^{\infty} Z(E)F(E)\, dE$ $\hspace{2cm}$ (2)

Were,

$$Z(E) = \frac{4\pi}{h^3} \left(2m_e^{*3/2}\right) (E - E_c^{1/2})\, dE$$

$$F(E) = \exp\left[E_F - E/KT\right]$$

$$\therefore\ n = \int_{E_C}^{\infty} \frac{4\pi}{h^3} \left(2m_e^{*3/2}\right) \left(E - E_c^{1/2}\right) \exp\left[E_F - E/KT\right] dE$$

$$n = \frac{4\pi}{h^3} \left(2m_e^{*3/2}\right) \int_{E_C}^{\infty} \left(E - E_c^{1/2}\right) \exp\left[E_F - E/KT\right] dE$$

$$n = \frac{4\pi}{h^3} \left(2m_e^{*3/2}\right) \exp\left[E_F/KT\right] \int_{E_C}^{\infty} \left(E - E_c^{1/2}\right) \exp\left[-E/KT\right]\ dE \hspace{2cm} (3)$$

To Solve (3),

When,	When,	When,
$E - E_c = x$	$E = E_c$	$E = \infty$
$\therefore E = E_c + x$	0	∞
$dE = dx$		

$$\therefore n = \frac{4\pi}{h^3}\left(2m_e^{*3/2}\right)\exp\left[E_F/KT\right]\int_0^\infty (x^{1/2})\exp\left[-(E_c + x)/KT\right]dx$$

$$\therefore n = \frac{4\pi}{h^3}\left(2m_e^{*3/2}\right)\exp\left[(E_F - E_c)/KT\right]\int_0^\infty (x^{1/2})\exp\left[-(x)/KT\right]dx$$

Using gamma function,

$$\int_0^\infty (x^{1/2})\exp\left[-(x)/KT\right]dx = (KT)^{3/2}\left(\frac{\pi^{1/2}}{2}\right)$$

$$\therefore n = \frac{4\pi}{h^3}\left(2m_e^{*3/2}\right)\exp\left[(E_F - E_c)/KT\right](KT)^{3/2}\left(\frac{\pi^{1/2}}{2}\right)$$

$$\therefore n = \frac{4\pi}{2h^3}(2\pi m_e^* KT)^{3/2}\exp\left[(E_F - E_c)/KT\right]$$

$$\therefore n = 2(2\pi m_e^* KT/h^2)^{3/2}\exp\left[(E_F - E_c)/KT\right] \tag{4}$$

This is the expression for carrier concentration of electrons in the conduction band of an intrinsic semiconductor.

2) Carrier Concentration of Holes in the Valence Band of an Intrinsic Semiconductor

Derivation

The No of holes in the energy level $\quad$ E & dE $\;$ (dp) = Z(E) [1 − F(E) dE] $\hspace{2cm}$ (1)

The No of holes in the valence band (p) = $\int_{-\infty}^{E_v} Z(E)[1 - F(E)\,dE]$ $\hspace{2cm}$ (2)

Were,

$$Z(E) = \frac{4\pi}{h^3}\left(2m_h^{*3/2}\right)(E_v - E^{1/2})\,dE$$

$$[1 - F(E)] = \exp\left[E - E_F/KT\right]$$

$$\therefore p = \int_{-\infty}^{E_v}\frac{4\pi}{h^3}\left(2m_h^{*3/2}\right)(E_v - E^{1/2})\exp\left[E - E_F/KT\right]dE$$

$$p = \frac{4\pi}{h^3}\left(2m_h^{*3/2}\right)\int_{-\infty}^{E_v}(E_v - E^{1/2})\exp\left[E - E_F/KT\right]dE$$

$$p = \frac{4\pi}{h^3}\left(2m_h^{*3/2}\right)\exp\left[-E_F/KT\right]\int_{-\infty}^{E_v}(E_v - E^{1/2})\exp\left[E/KT\right]dE \tag{3}$$

To Solve (3),

When,	When,	When,
$E_v - E = x$	$E = E_v$	$E = -\infty$
$\therefore E = E_v - x$	0	$-\infty$
$dE = -dx$		

$$\therefore p = \frac{4\pi}{h^3} \left(2m_h^{*\,3/2}\right) \exp\left[-E_F/KT\right] \int_\infty^0 (x^{1/2}) \exp\left[(E_v - x)/KT\right](-dx)$$

$$\therefore p = \frac{4\pi}{h^3} \left(2m_h^{*\,3/2}\right) \exp\left[(E_v - E_F)/KT\right] \int_0^\infty (x^{1/2}) \exp\left[-(x)/KT\right] dx$$

Using gamma function,

$$\int_0^\infty (x^{1/2}) \exp\left[-(x)/KT\right] dx = (KT)^{3/2} \left(\frac{\pi^{1/2}}{2}\right)$$

$$\therefore p = \frac{4\pi}{h^3} \left(2m_h^{*\,3/2}\right) \exp\left[(E_v - E_F)/KT\right] (KT)^{3/2} \left(\frac{\pi^{1/2}}{2}\right)$$

$$\therefore p = \frac{4\pi}{2h^3} (2\pi m_h^* KT)^{3/2} \exp\left[(E_v - E_F)/KT\right]$$

$$\therefore p = 2 (2\pi m_h^* KT/h^2)^{3/2} \exp\left[(E_v - E_F)/KT\right] \tag{4}$$

This is the expression for carrier concentration of holes in the valence band of an intrinsic semiconductor.

3) *Expression for Intrinsic Carrier Concentration in an Intrinsic Semiconductors*

In an intrinsic semiconductor, The concentration of electrons in CB is equal to the concentration of holes in the valence band.

$$n_i^2 = 4 \left(\frac{2\pi KT}{h^2}\right)^3 (m_e^* m_h^*)^{\,3/2} \exp\left[(E_v - E_c)/KT\right]$$

$$n_i^2 = 4 \left(\frac{2\pi KT}{h^2}\right)^3 (m_e^* m_h^*)^{\,3/2} \exp\left[-E_g/KT\right]$$

Where, $(E_c - E_v) = E_g$ is the forbidden energy gap.

$$\therefore n_i^2 = 2 \left(\frac{2\pi KT}{h^2}\right)^{3/2} (m_e^* m_h^*)^{\,3/4} \exp\left[-E_g/2KT\right]$$

4) *Explain the Variation of FERMI Level with Temperature in an Intrinsic Semiconductor*

In an intrinsic semi conductor, the No of holes in the valence band is equal to the No of electrons in the conduction band

$$n = p$$

$$2(2\pi m_e^* KT/h^2)^{3/2} \exp[(E_F - E_c)/KT] = 2 (2\pi m_h^* KT/h^2)^{3/2} \exp[(E_v - E_F)/KT]$$

(or)

$$(m_e^*)^{3/2} \exp[(E_F - E_c)/KT] = (m_h^*)^{3/2} \exp[(E_v - E_F)/KT] \quad (or)$$

$$\exp[2E_F/KT] = (m_h^*/m_e^*)^{3/2} \exp[(E_v + E_c)/KT]$$

Taking log on both sides

$$2E_F/KT = 3/2 (m_h^*/m_e^*) + (E_v + E_c)/KT$$

$$E_F = 3KT/4 \log(m_h^*/m_e^*) + (E_v + E_c)/2 \qquad (1)$$

Assuming that, $m_e^* = m_h^*$

$$\log(m_h^*/m_e^*) = \log 1 = 0 ; \text{ If } T=0K;$$

∴ Equation (1) becomes,

$$E_F = (E_v + E_c)/2$$

Thus, the Fermi level is located half way between the top of valence band and bottom the conduction band as in figure (a). Fermi level rises slightly with increasing temperature as in figure (b).

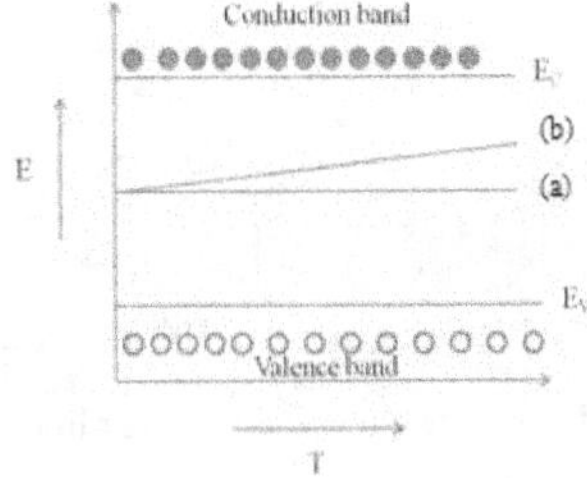

Figure 2.3: Representation of Fermi Level Intrinsic Semiconductor

5) The Energy Gap of a Semiconductor and its Experimental Verification

The electrical conductivity of an intrinsic semiconductor is given by,

$$\sigma i = A \, e^{-(Eg/2KT)}$$

Where,

$$A = 2e\left(\frac{2\pi KT}{h^2}\right)^{3/2} (m_e^* m_h^*)^{3/4} (\mu_e + \mu_h)$$

∴ Resistivity, $\rho_i = 1/\sigma_i$

$$= 1/A \, e^{-(Eg/2KT)}$$

$$= e^{(Eg/2KT)}/A$$

$$[\rho_i = R_i a/l]$$

$$\therefore R_i a/l = e^{(Eg/2KT)} / A$$

$$R_i = l\, e^{(Eg/2KT)} /aA$$

$$R_i = C\, e^{(Eg/2KT)}$$

$$\text{Since, } C = l/aA$$

Taking log on both sides,

$$\text{Log } R_i = \log (C\, e^{(Eg/2KT)})$$

$$\text{Log } R_i = \log C + \log e^{(Eg/2KT)}$$

$$\text{Log } R_i = \log C + Eg/2KT$$

Equation (1) is similar to the equation of a straight line,

$$Y = mx + C$$

$$Y = \log R_i \;; C = \log C; m = Eg/2K \;; x = 1/T$$

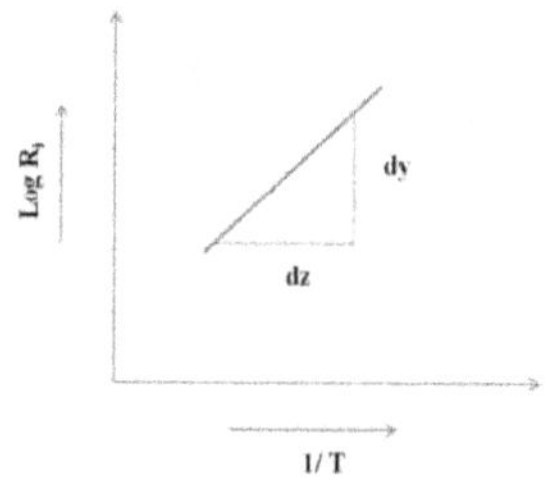

Figure 2.4: Graphical Representation of Determination of Band Gap

The value of E_g is determined from the slope of a straight line.

$$E_g = 2K\, dy/dx$$

Experimental determination of band gap:

1) A Semiconducting material is connected in series with a battery, rheostat and an ammeter.
2) It is immersed in an oil bath.
3) A voltmeter is connected in parallel to the specimen.
4) The temperature is noted.
5) Known potential difference V is applied across the specimen and the current is noted.
6) The potential difference is kept constant.
7) The experiment is repeated for at different temperatures.
8) The graph is drawn between log R_i and 1/T. The slope of the straight line is determined.
9) The energy gao is determined by the relation,

$$E_g = 2K \times slope$$

2.2.2. Carrier Concentration of Electrons in N-type Semi Conductors

N – Type Semiconductor

1) When a small amount of pentavalent impurity (group V element) is added to a pure semiconductor, it becomes a n – semiconductor.

2) Such impurities are known as donor impurities because they donate the free electrons to the semiconductor crystal.

3) A pentavalent impurity (arsenic) having five valance electrons is added to a pure semiconducting material having four valance electrons (silicon or germanium).

4) The four valance electrons of the impurity atoms bond with four valance electrons of the semiconductor atom and remaining 1 electron of the impurity atom is left free as shown figure 2.5.

5) Therefore number of free electrons increases, as the electrons are produced in excess, they the majority charge carrier in n – type semiconductor and holes are the majority charge carriers.

Since electrons are donated in this type of semiconductor the energy level of these donated electrons is called donor energy level (Ed) as shown in fig.

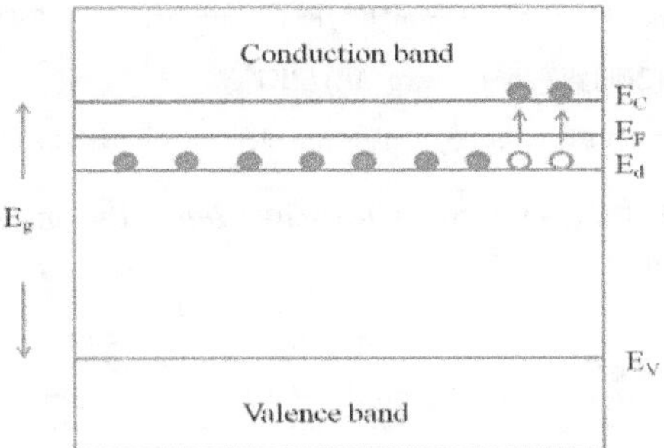

Figure 2.5: The Energy Band Diagram of a n-type Semiconductor

Density of electrons per unit volume in the CB,

$$n = 2(2\pi m_e^* KT/h^2)^{3/2} \exp\left[(E_F - E_c)/KT\right] \tag{1}$$

Density of ionised donors, $Nd^+ = N_d / (1 + \exp\left[(E_F - E_d)/KT\right]$ (2)

Since, (1) = (2)

$$2(2\pi m_e^* KT/h^2)^{3/2} \exp\left[(E_F - E_c)/KT\right] = N_d / (1 + \exp\left[(E_F - E_d)/KT\right]$$

$$2(2\pi m_e^* KT/h^2)^{3/2} \exp\left[(E_F - E_c)/KT\right] = N_d / (\exp\left[(E_F - E_d)/KT\right]$$

$$2(2\pi m_e^* KT/h^2)^{3/2} \exp\left[(E_F - E_c)/KT\right] \quad = N_d \exp\left[(E_d - E_F)/KT)\right] \tag{3}$$

Taking log on both sides,

$$\log 2(2\pi m_e^* KT/h^2)^{3/2} + \log \exp\left[(E_F - E_c)/KT\right] \quad = \log N_d \exp\left[(E_d - E_F)/KT)\right]$$

$$\log 2(2\pi m_e^* KT/h^2)^{3/2} + \left[(E_F - E_c)/KT\right] \quad = \log N_d + \left[(E_d - E_F)/KT)\right]$$

Rearranging we have,

$$(E_F - E_c - E_d + E_F)/\,KT = \log N_d - \log\left[2(2\pi m_e^* KT/h^2)^{3/2}\right]$$

$$(2E_F - E_c - E_d)/\,KT = \log\left[N_d / [2(2\pi m_e^* KT/h^2)^{3/2}]\right]$$

$$2E_F = E_d + E_c + KT \log\left[N_d / [2(2\pi m_e^* KT/h^2)^{3/2}]\right]$$

$$E_F = \left(\frac{E_d + E_c}{2}\right) + \frac{KT}{2} \log\left[N_d / [2(2\pi m_e^* KT/h^2)^{3/2}]\right]$$

$$E_F = \left(\frac{E_d + E_c}{2KT}\right) + \frac{1}{2} \log\left[N_d / [2(2\pi m_e^* KT/h^2)^{3/2}]\right]$$

$$E_F = \left(\frac{E_d + E_c}{2KT}\right) + \log\left[N_d{}^{1/2} / [2(2\pi m_e^* KT/h^2)^{3/2}]\right]^{1/2} \tag{4}$$

Substitute (4) in (1);

$$n = 2 \left(\frac{2\pi m_e^* KT}{h^2}\right)^{3/2} \exp\left[(E_d - E_c)/2KT) + \log\left[N_d{}^{1/2} / [2(2\pi m_e^* KT/h^2)^{3/2}]^{1/2}\right]\right]$$

$$n = 2 \left(\frac{2\pi m_e^* KT}{h^2}\right)^{3/2} \left[\left[N_d{}^{1/2} / [2(2\pi m_e^* KT/h^2)^{3/2}]^{1/2}\right]\right] \exp(E_d - E_c)/2KT$$

$$n = 2^{1/2}\, N_d{}^{1/2}\, 2(2\pi m_e^* KT/h^2)^{3/4} \exp(E_d - E_c)/2KT$$

$$n = (2N_d)^{1/2}\, 2(2\pi m_e^* KT/h^2)^{3/4} \exp(\Delta E)/2KT \tag{5}$$

This is the expression for carrier concentration electrons in N-type semiconductors.

Variation of Fermi Level with Temperature and Impurity Concentration

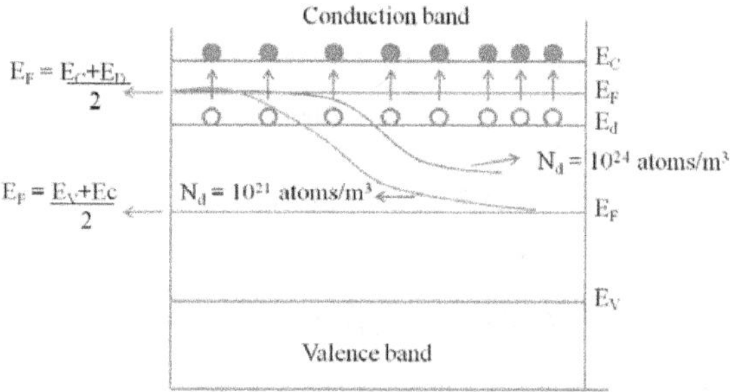

Figure 2.6: Energy Band Diagram of N-Type Semiconductor

1) As shown in figure 2.6, at 0K, Fermi level lies exactly in the middle of donor level (E_d) and bottom of the conduction band.

$$E_F = \frac{E_d + E_c}{2}$$

2) When temperature is increased, some of the electrons in E_d level may be shifted to conduction band. Hence, some vacant sites are created in E_d levels. Therefore, the Fermi level shifts down to separate that empty levels and the filled valence band level. Hence, reaches intrinsic behaviour.

3) For the same temperature, if the impurity atoms increased, the electron concentration gets increased and hence Fermi level increases.

2.2.3. *Carrier Concentration of Holes in P-type Semi Conductors*

P – Type Semiconductor

1) P–type semiconductor is obtained by doping an intrinsic semiconductor with trivalannt (3 electrons in valance band) impurity atoms like boron, gallium, indium etc.,

2) The three valance electrons of the impurity atom pairs with three valence electrons of the semiconductor atom and one position of the impurity atom remains vacant, this is called hole as shown in figure 2.7.Therefore the number of holes increased with impurity atoms added to it. Since holes are produced in excess, they are the majority charge carriers in p–type semiconductor and electrons are the minority charge carriers.

3) Since the impurity can accept the electrons this energy level is called acceptor energy level (Ea) and is present just above the valence band as shown in figure 2.7.

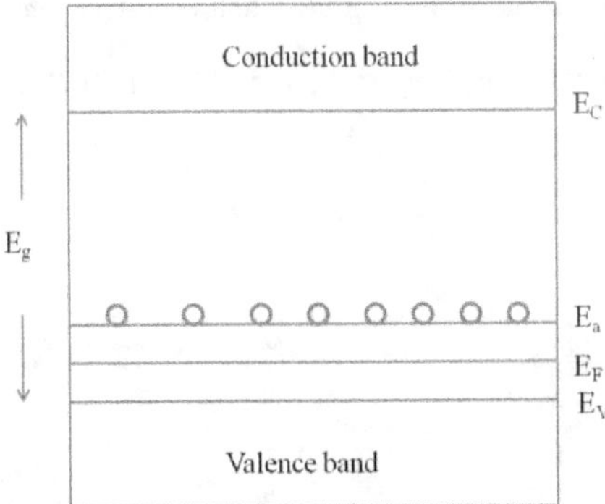

Figure 2.7: The Energy Band Diagram of a p-Type Semiconductor

Density of holes per unit volume in the VB,

$$p = 2(2\pi m_h^* KT/h^2)^{3/2} \exp\left[(E_v - E_F)/KT\right] \qquad (1)$$

Density of ionised donors, $N_a^+ = N_a \exp\left[(E_F - E_a)/KT\right]$ \qquad (2)

Since, (1) = (2)

$$2(2\pi m_h^* KT/h^2)^{3/2} \exp[(E_V - E_F)/KT] = N_a \exp[(E_F - E_a)/KT]$$

Taking log on both sides,

$$\log 2(2\pi m_h^* KT/h^2)^{3/2} \exp[(E_V - E_F)/KT] = \log N_a \exp[(E_F - E_a)/KT]$$

$$\log 2(2\pi m_h^* KT/h^2)^{3/2} + [(E_V - E_F)/KT] = \log N_a + [(E_F - E_a)/KT)]$$

Rearranging we have,

$$(E_F - E_a - E_v + E_F)/2 = -\log N_a + \log[2(2\pi m_h^* KT/h^2)^{3/2}]$$

$$(2E_F - (E_a + E_V))/KT = -\log[N_a / [2(2\pi m_h^* KT/h^2)^{3/2}]]$$

$$2E_F = (E_a + E_V) - KT \log[N_a / [2(2\pi m_h^* KT/h^2)^{3/2}]]$$

$$E_F = \left(\frac{E_a + E_V}{2}\right) - \frac{KT}{2} \log[N_a / [2(2\pi m_h^* KT/h^2)^{3/2}]]$$

$$E_F = \left(\frac{E_a + E_V}{2KT}\right) - \frac{1}{2} \log[N_a / [2(2\pi m_h^* KT/h^2)^{3/2}]] \qquad (4)$$

Substitute (4) in (1);

$$p = 2\left(\frac{2\pi m_h^* KT}{h^2}\right)^{3/2} \exp\left[(2E_V - E_v - E_a)/2KT) + \frac{1}{2}\log[N_a / [2(2\pi m_h^* KT/h^2)^{3/2}]\right]$$

$$p = 2\left(\frac{2\pi m_h^* KT}{h^2}\right)^{3/2} \left[[N_a^{1/2} / [2(2\pi m_h^* KT/h^2)^{3/2}]^{1/2}]\right]$$

$$n = 2^{1/2} N_a^{1/2} \, 2(2\pi m_h^* KT/h^2)^{3/4} \exp(E_v - E_a)/2KT$$

$$n = (2N_a)^{1/2} \, 2(2\pi m_h^* KT/h^2)^{3/4} \exp -(\Delta E)/2KT \qquad (5)$$

This is the expression for carrier concentration holes in P-type semiconductors.

Variation of Fermi Level with Temperature and Impurity Concentration

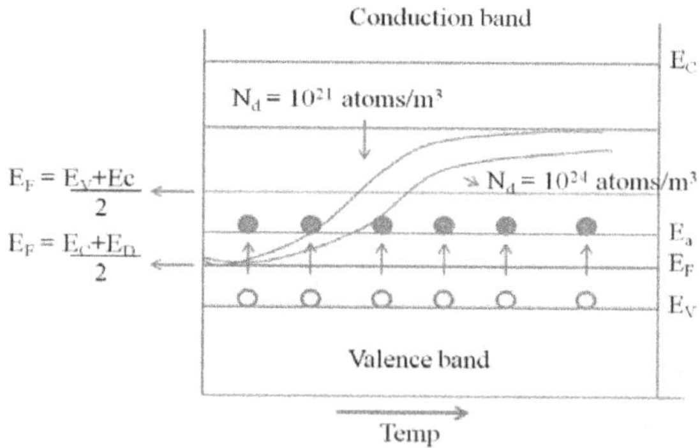

Figure 2.8: The Energy Band Diagram of a p-Type Semiconductor

1) As shown in figure 2.8, at 0K, Fermi level lies exactly half way between acceptor level E_a and top of the valence band E_v.

$$E_F = \frac{E_a + E_v}{2}$$

2) When temperature is increased, some of the electrons in the valence band will go to acceptor energy levels by breaking up the covalent bonds. Hence, Fermi level is shifted in upward direction.

3) For the same temperature, if the impurity atoms increased, the hole concentration increases and hence the Fermi level decreases.

2.3.　Hall Effect

Measurement of conductivity will not determine whether the conduction is due to electron or holes and therefore it will be very difficult to distinguish between p–type and n-type semiconductors.

Therefore Hall Effect is used to distinguish between the two types of charge carriers and their carrier densities and is used to determine the mobility of charge carriers.

Principle

When a conductor (metal or semiconductor) carrying current (I) is placed perpendicular to a magnetic field (B), a potential difference (electric field) is developed inside the conductor in a direction perpendicular to both current and magnetic field. This phenomenon is known as Hall Effect and the voltage thus generated is called Hall voltage.

THEORY

1) Hall Effect in n- type Semiconductor

Let us consider a n-type semiconductor material in the form of rectangular slab. In such a material current flows in X – direction and magnetic field B applied in Z- direction. As a result, Hall voltage is developed along Y – direction as shown in figure 2.9.

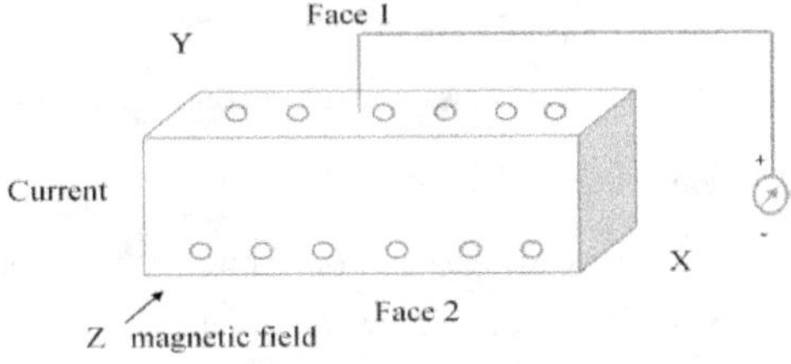

Figure 2.9: Representation of Hall Effect in N-type Semiconductor

Since the direction of current is from left to right the electrons moves from right to left in X-direction as shown in figure 2.9.1. Now due to magnetic field applied the electron moves towards downward direction with velocity v and cause negative charge to accumulate at face (1) of the material as shown in fig. therefore the potential difference is established between the face (2) and face (1) of the specimen which gives rise to EH in the negative Y direction.

Face 1

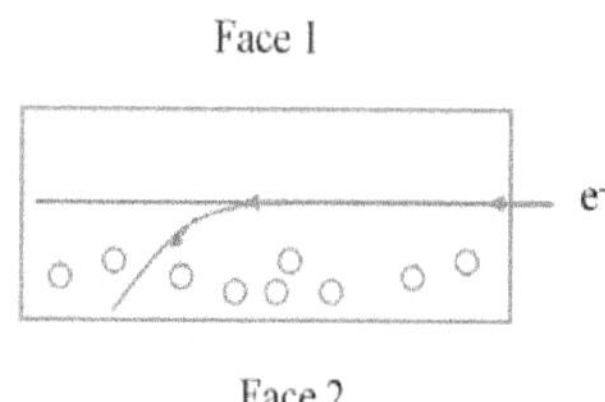

Face 2

Figure 2.9.1: Direction of Current Due to the Applied Field

Here, the force due to potential difference = -e EH   (1)

Force due magnetic field = - Bev   (2)

At equilibrium equation (1) = equation (2)

$$-e\ EH = - Bev$$
$$EH = Bv \qquad (3)$$

We know the current density J_x in the X- direction is

$$J_x = -n_e\ ev$$
$$v = - J_x / n_e\ e \qquad (4)$$

Substituting equation (4) in equation (3) we get

$$E_H = - B\ Jx / n_e\ e \qquad (5)$$
$$E_H = R_H . J_x . B \qquad (6)$$

Where R_H is known as the Hall co – efficient, is given by

$$R_H = - (1 / n_e\ e) \qquad (7)$$

The negative sign indicates that the field is developed in the negative Y – direction.

2) *Hall Effect in p – type Semiconductor*

Let us consider a p – type semiconducting material for which the current is passed along X – direction from left to right and magnetic field is applied along Z – direction as shown in figure 3.1. Since the direction of current is from left to right, the holes will also move in the same direction as shown in figure 3.2.

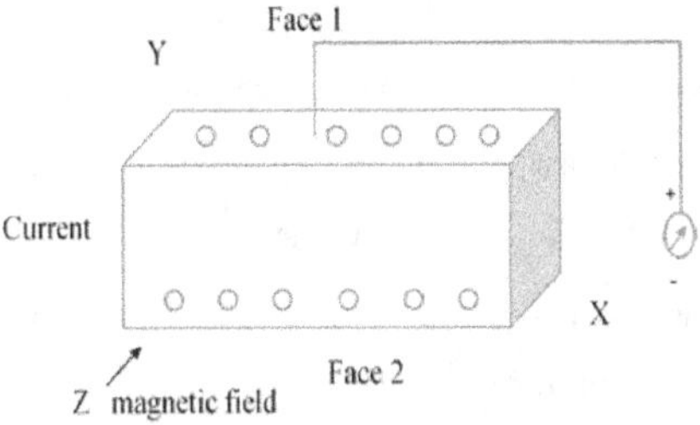

Figure 2.9.2: Representation of Hall Effect in P-type Semiconductor

Now due to magnetic field applied the holes moves towards downward direction with velocity v and accumulates at the face (1). A potential difference is established between face (1) and (2) in the positive Y - direction.

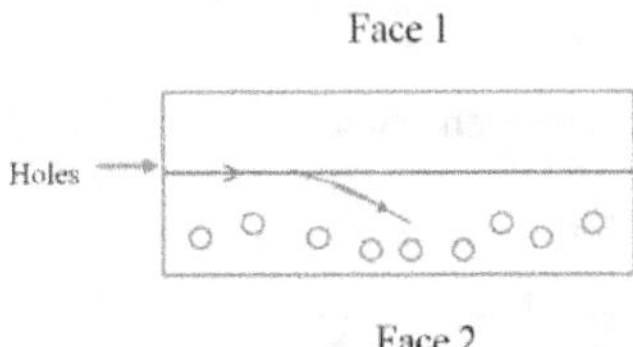

Figure 2.9.3: Direction of Current due to the Applied Field

Here, the force due to potential difference = -e EH   (8)

Force due magnetic field = Bev   (9)

At equilibrium equation (1) = equation (2)

$$e\ E_H = Bev$$

$$E_H = Bv \qquad (10)$$

We know the current density J_x in the X- direction is

$$J_x = n_h\ ev$$

$$v = J_x / n_h e \qquad (11)$$

Substituting equation (11) in equation (10) we get

$$E_H = B\ J_x / n_h e \ ; \ E_H = R_H. J_x. B$$

Where RH is known as the Hall co–efficient, is given by

$$R_H = (1 / n_h\ e) \qquad (12)$$

The positive sign indicates that the field is developed in the positive Y – direction

3) *Hall Coefficient in Terms of Hall Voltage*

If the thickness of the sample is t and the voltage developed is VH,

$$\text{then Hall voltage } V_H = E_H \cdot t \qquad (13)$$

Substituting equation (6) in equation (13), we have

$$V_H = R_H \, J_x \, B \cdot t \qquad (14)$$

If b is the width of the sample then

Area of the sample = b.t

$$\text{Current density} = J_x = I_x / bt \qquad (15)$$

Substituting equation (15) in equation (14) we get, $V_H = \dfrac{R_H \, I_x \, B \, t}{bt}$

$$V_H = \dfrac{R_H \, I_x \, B}{b}$$

$$\text{Hall coefficient, } R_H = \dfrac{R_H \, I_x \, B}{b} \qquad (16)$$

The sign for VH will be opposite for n and p type semiconductors.

4) *Mobility of Charge Carriers*

We know that Hall coefficient,

$$R_H = - \, (1 \, / \, n_e \, e)$$

The above expression is valid only for conductors where the velocity is taken as the drift velocity. But for semiconductor velocity is taken as average velocity so RH for n- type semiconductor is modified as

$$R_H = - \, 3\pi \, / \, 8[1/n_e \, e]$$

$$R_H = - \, 1.18 \, / \, ne \, e \qquad (1)$$

We know the conductivity for n – type is $\sigma e = n_e \, e \, \mu_e$

$$\text{Or} \quad \mu_e = \sigma_e / n_e \, e \qquad (2)$$

From equation (1) we can write

$$1 \, / \, n_e e = - \, R_H \, / \, 1.18 \qquad (3)$$

Substituting equation (3) in equation (2) we get

$$\mu_e = - \, [\, \sigma e \, R_H \, / \, 1.18 \,] \quad \text{We know that, } R_H = \dfrac{V_H \, B}{I_x \, B}$$

For n – type semiconductor, the mobility of electron is

$$\mu e = - \, [\, \sigma_e / 1.18 \,] \left(\dfrac{V_H \, B}{I_x \, B} \right) \qquad (4)$$

Similarly for p – type semiconductor, the mobility of hole is

$$\mu_h = [\, \sigma_e / 1.18 \,]\left(\frac{V_H B}{I_x B}\right) \qquad (5)$$

5) Experimental Determination of Hall Effect

A semiconducting material is taken in the form of a rectangular slab of thickness t and breadth. A suitable current I_x ampere is passed through this sample along X- axis by connecting it to a battery.

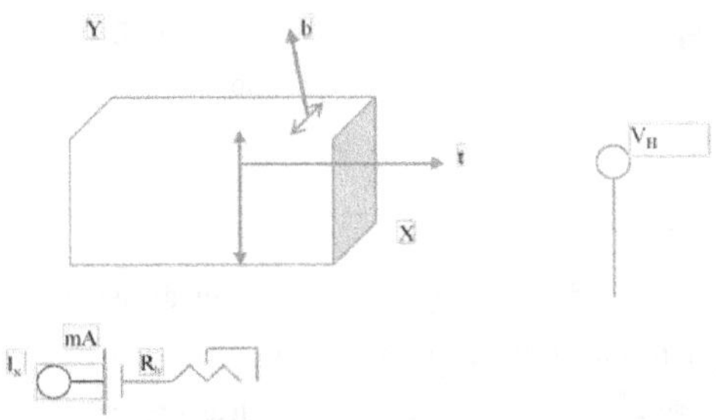

Figure 2.9.4: Experimental Determination of Hall Effect

This sample is placed in between two poles of an electro magnet such that magnetic field is applied along Z – axis.

Due to Hall Effect, Hall voltage (V_H) is developed in the sample. This voltage measured by fixing two probes at the centers of the bottom and top faces faces of the sample. By measuring Hall voltage, Hall coefficient is determined from the formula

$$\text{Hall coefficient, } R_H = \frac{V_H B}{I_x B}$$

From the hall coefficient, carrier concentration and mobility can be determined.

6) Applications of Hall Effect

1) It is used to determine whether the material is p-type or n-type semiconductor.(ie) if R_H is negative then the material n- type. If the R_H is positive then the material p-type.
2) It is used to find the carrier concentration.
3) It is used to find the mobility of charge carriers μ_e, μ_h.
4) It is used to find the sign of the current carrying charges.
5) It is used to design magnetic flux meters and multipliers on the basis of Hall voltage.
6) It is used to find the power flow in an electromagnetic wave.

Unit-III

Magnetic Materials and Superconducting Materials

Abstract

After studying this unit, one can acquire

1) Understand about dia, para and ferro magnetism.
2) Acquire knowledge on applications of ferrites.
3) Acquire proper knowledge on dielectrics and their properties.
4) Understand the concepts of dielectric breakdown.
5) Understand about the nature, properties and applications of superconductors.

3.1.　Introduction

Magnetic materials are the materials which can be made to behave as magnets. When these materials are kept in an external magnetic field, they will create a permanent magnetic moment in it. A very large number of modern devices depend upon magnetic properties of materials for their working. For example, the speakers, electrical power generators, electrical machines, transformers, television, data storage devices like magnetic tapes and disks, magnetic compass etc. The MRI field.

An understanding of the behaviour of magnetic materials will be helpful not only in the selection of suitable materials for a particular application but also in designing new applications of these materials.

Among the different eleven types of magnetic materials, only five magnetic materials are the most important for the practical application. They are:

1) Diamagnetic materials
2) Paramagnetic materials
3) Ferromagnetic materials
4) Anti ferromagnetic materials
5) Ferrimagnetic materials or ferrites

3.2.　Basic Terms and Definitions

1) Magnetic Flux (φ)

Total number of magnetic lines of force passing through a surface is known as magnetic flux. It is represented by the symbol 'φ' and its unit is weber.

2) *Magnetic Flux Density (or) Magnetic Induction (B)*

Magnetic flux density at any point in a magnetic field is defined as the magnetic flux (φ) passing normally through unit area of cross section (A) at that point. It is denoted by the symbol B and its unit is weber/metre^2 or tesla.

$$B = [\varphi / A]$$

3) *Intensity of Magnetization (I)*

1) The term magnetization means the process of converting non-magnetic material on magnetic material.

2) When some amount of external magnetic field is applied to the metals such as iron, steel and alloys etc., they are magnetized to different degrees. The intensity of magnetisation (I) is the measure of the magnetisation of a magnetized specimen. It is defined as the magnetic moment per unit volume.

$$I = M / V \quad (weber / metre^2)$$

4) *Magnetic Field Intensity (or) Strength (H)*

Magnetic field intensity at any point in a magnetic field is the force experienced by unit north pole placed at that point.

It is denoted by H and its unit is Newton per weber or ampere turns per metre (A/m).

5) *Magnetic Permeability (μ)*

Magnetic permeability of a substance measure the degree to which the magnetic field can penetrate through the substance.

It is found that magnetic flux density (B) is directly proportional to the magnetic field strength (H).

$$B \, \alpha \, H$$
$$B = \mu \, H$$

Where is a constant of proportionality and it is known as permeability or absolute permeability of the medium.

$$\mu = B / H$$

Hence, the permeability of a substance is the ratio of the magnetic flux density (B) inside the substance to the magnetic field intensity (H).

6) Absolute Permeability

Absolute permeability of a medium or material is defined as the product of permeability of free space (μ_0) and the relative permeability of the medium (μ_r).

$$\mu = \mu_0 \times \mu_r$$

7) Relative Permeability (μr) of a Medium

Relative permeability of a medium is defined as the ratio between absolute permeability of a medium (μ) to the permeability of a free space

$$(\mu_0) \quad \mu_r = \mu / \mu_0$$

8) Magnetic Susceptibility (χ)

Magnetic susceptibility (χ) of a specimen is a measure of how easily a specimen can be magnetized in a magnetic field.

It is the ratio of intensity of magnetisation (I) induced in it to the magnetizing field (H).

$$\chi = I / H$$

9) Retentivity (or) Remanence

When the external magnetic field is applied to a magnetic material is removed, the magnetic material will not loss its magnetic property immediately. There exits some residual intensity of magnetization in the specimen even when the magnetic field is cut off. This is called residual magnetism (or) retentivity.

10) Coercivity

The residual magnetism can be completely removed from the material by applying a reverse magnetic field. Hence coercivity of the magnetic material is the strength of reverse magnetic field (- Hc) which is used to completely demagnetize the material.

11) Origin of Magnetic Moment and Bohr Magneton

1) Origin of Magnetic Moment

1) Any matter is basically made up of atoms. The property of magnetism exhibited by certain materials with the magnetic property of its constituent atoms. We know that electrons in an atom revolve around the nucleus in different orbits.
2) Basically there are three contributions for the magnetic dipole moment of an atom.

3) The orbital motions of electrons (the motion of electrons in the closed orbits around the nucleus) are called orbital magnetic moment.

4) Spin motion of the electrons (due to electron spin angular momentum) is called spin magnetic moment.

5) The contribution from the nuclear spin (due to nuclear spin angular momentum) is nearly 10^{-3} times smaller than that of electron spin; it is not taken into consideration.

2) Bohr Magneton

The magnetic moment contributed by an electron with angular momentum quantum number n = 1 is known as Bohr Magneton.

3.3. Different Types of Magnetic Materials

1) Diamagnetic Materials

Diamagnetism is exhibited by all the materials. The atoms in the diamagnetic materials do not possess permanent magnetic moment. However, when a material is placed in a magnetic field, the electrons in the atomic orbits tend to counteract the external magnetic field and the atoms acquire an induced magnetic moment. As a result, the material becomes magnetized. The direction of the induced dipole moment is opposite to that of externally applied magnetic field. Due to this effect, the material gets very weakly repelled, in the magnetic field. This phenomenon is known as diamagnetism. When a magnetic field Ho is applied in the direction shown in figure 3.3.1, the atoms acquire an induced magnetic moment in the opposite direction to that of the field. The strength of the induced magnetic moment is proportional to the applied field and hence magnetization of the material varies directly with the strength of the magnetic field. The induced dipoles and magnetization vanish as soon as the applied field is removed.

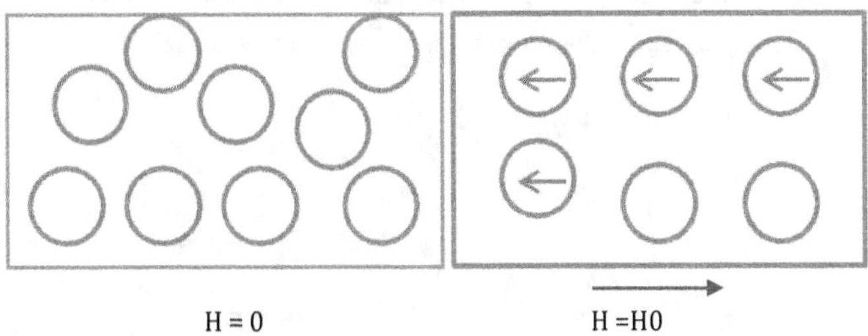

Figure 3.3.1: Representation of Dia Magnetic Material in the Absence and Presence of Magnetic Field

Properties of Diamagnetic Material

1) Diamagnetic magnetic material repels the magnetic lines of force. The behaviour of diamagnetic material in the presence of magnetic field is shown in figure 3.3.2.

2) There is no permanent dipole moment. Therefore, the magnetic effects are very small.

3) The magnetic susceptibility is negative and it is independent of temperature and applied magnetic field strength.

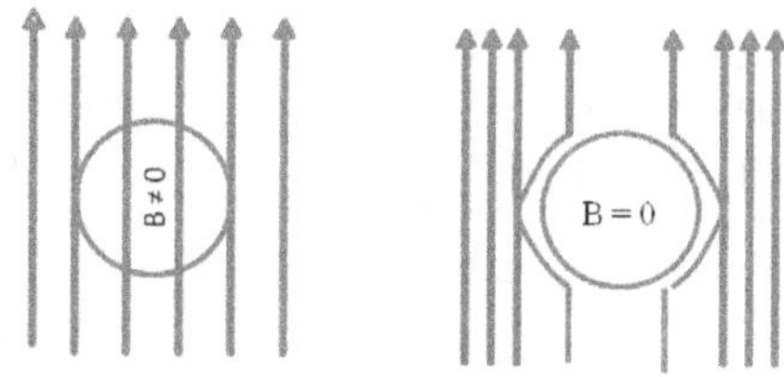

Figure 3.3.2: Representation of Dia Magnetic Material in Presence of High and Low Magnetic Fields

2) Paramagnetic Materials

1) In certain materials, each atom or molecule possesses a net permanent magnetic moment (due to orbital and spin magnetic moment) even in the absence of an external magnetic field.

2) The magnetic moments are randomly oriented in the absence of external magnetic field.

3) Therefore the net magnetic moment is zero, and hence the magnetization of the material is zero.

4) But, when an external magnetic field is applied, the magnetic dipoles tend to align themselves in the direction of the magnetic field and the material becomes magnetized. As shown in figure 3.3.3. This effect is known as paramagnetism.

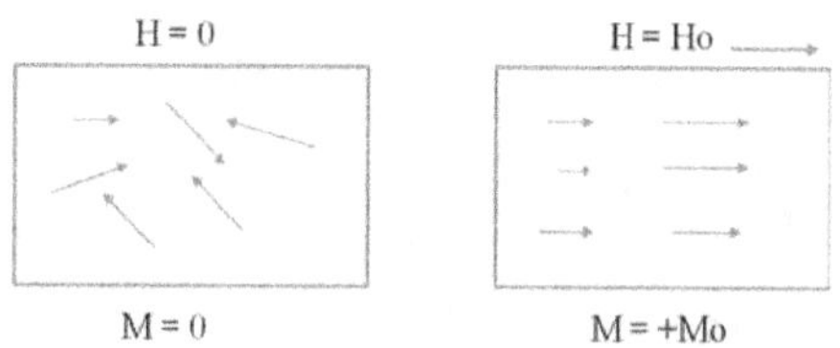

Figure 3.3.3: Representation of Dia Magnetic Material in the Absence and Presence of Magnetic Field

Thermal agitation disturbs the alignment of the magnetic moments. With an increase in temperature, the increase in thermal agitation tends to randomize the dipole direction thus leading to a decrease in magnetization. This indicates that the paramagnetic susceptibility decreases with increases in temperature. It is noted that the paramagnetic susceptibility varies inversely with temperature.

$$\chi \alpha \, 1/T \; ; \; \chi = C/T$$

This is known as Curie law of paramagnetism and C is a constant called Curie constant.

Properties of Paramagnetic Materials

Paramagnetic materials attract magnetic lines of force. They possess permanent dipole moment. The susceptibility is positive and depend on temperature is given by $\chi = C/T - \theta$. The spin alignment is shown in figure below.

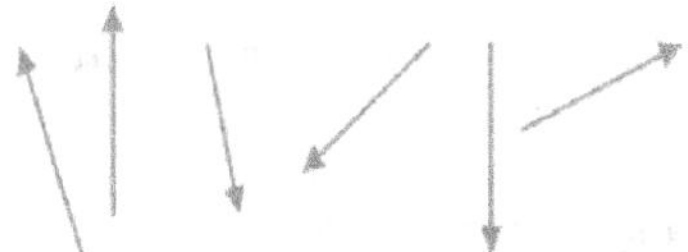

Example - Manganous sulphate, ferric oxide, ferrous sulphate, nickel sulphate, etc.

3) Ferromagnetic Materials

Certain materials like iron, cobalt, nickel and certain alloys exhibit high degree of magnetization. These materials show spontaneous magnetization. (i.e) they have small amount of magnetization even in the absence of external magnetic field. This indicates that there is strong internal field within the material which makes atomic magnetic moments with each other. This phenomenon is known as ferromagnetism.

Properties of Ferromagnetic Materials

1) All the dipoles are aligned parallel to each other due to the magnetic interaction between the two dipoles.

2) They have permanent dipole moment. They are strongly attracted by the magnetic field.

3) They exhibit magnetization even in the absence of magnetic field. This property of ferromagnetic material is called as spontaneous magnetization.

4) They exhibit hysteresis curve. On heating, they lose their magnetization slowly. The dipole alignment is shown in figure below.

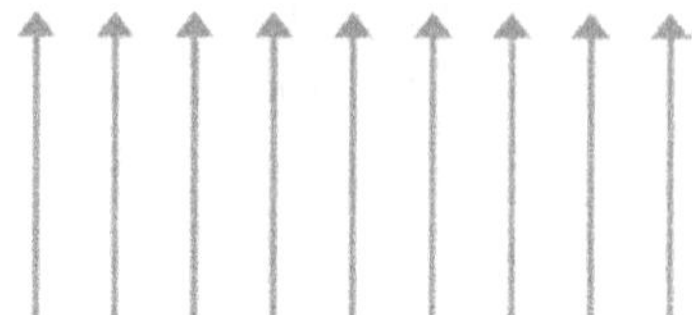

The susceptibility is very high and depends on the temperature. It is given by $\chi = C / T - \theta$ [for $T > \theta$; paramagnetic behaviour; for $T < \theta$; ferromagnetic behavior] Where C is the Curie constant and θ is the paramagnetic Curie temperature.

4) Anti Ferromagnetic Materials

1) Anti ferromagnetic materials are magnetic materials which exhibit a small positive susceptibility of the order of 10^{-3} to 10^{-5}.

2) In anti ferromagnetic materials, the susceptibility increases with increasing temperature and it reaches maximum at a certain temperature called Neel Temperature, T_N.

Properties of Anti Ferromagnetic Materials

1) The electron spin of neighboring atoms are aligned anti parallel. (i.e) the spin alignment is anti parallel as shown in below figure.

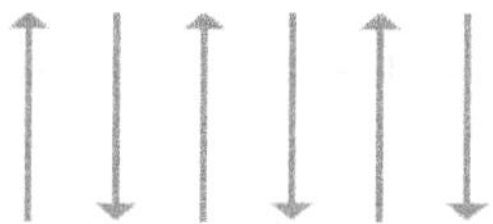

2) Anti ferromagnetic susceptibility is mainly depends on temperature.

The susceptibility of the anti ferromagneitc material is small and positive. It is given by

$$\chi = C / T + \theta \qquad \text{when } T > TN$$
$$\chi \propto T \qquad \text{when } T < TN$$

3) The susceptibility initially increases slightly with the temperature and beyond Neel temperature, the susceptibility decreases with temperature.

5) Ferrimagnetic Materials

Properties of Ferrites

1) Ferrites have net magnetic moment.

2) Above Curie temperature, it becomes paramagnetic, while it behaves ferromagnetic material blow Curie temperature.

3) The susceptibility of ferrite is very large and positive. It is temperature dependent and is given by

$$\chi = C / T \pm \theta \qquad \text{for } T > T_N$$

4) Spin alignment is antiparallel of different magnitudes as shown in below figure.

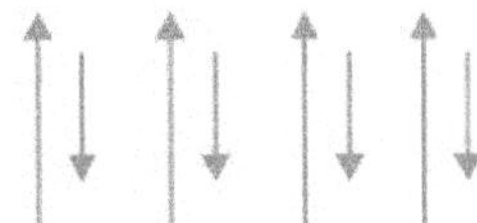

5) Mechanically it has pure iron character.

6) They have high permeability and resistivity.

7) They have low eddy current losses and low hysteresis losses.

6) Ferromagnetism

The materials which have finite value of magnetization even if the external magnetic field is absent are called ferromagnetic materials. This phenomenon is called ferromagnetism. The ferromagnetic materials exhibit high degree of magnetization.

a) Explanation

In a ferromagnetic material, the magnetic interactions between any two dipoles align themselves parallel to each other. Ferromagnetism arises due to the special form of interaction called exchange coupling between adjacent atoms. This exchange coupling is favourable for spin alignment and they coupling their magnetic moments together in rigid parallelism.

A ferromagnetic materials exibits ferromagnetic property below a particular temperature called ferromagnetic. Curie temperature (f_θ). Above f_θ they behave as paramagnetic material.

b) Domain Theory of Ferromagnetism

We can observe that a ferromagnetic material such as iron does not have magnetization unless they have been previously placed in an external magnetic field. But according to Weiss theory, the molecular magnets in the ferromagnetic material are said to be aligned in such way that, they exhibit magnetization even in the absence of external magnetic field. This is called spontaneous magnetization. (i.e) it should have some internal magnetization due to quantum exchange energy.

According to Weiss hypothesis, a single crystal of ferromagnetic material is divided into large number of small regions called domains. These domains have spontaneous magnetization due to the parallel alignment of spin magnetic moments in each atom. But the direction of spontaneous magnetization varies from domain to domain and is oriented in such way that the net magnetization of the specimen is zero.

The boundaries separating the domains are called domain walls. These domain walls are analogous to the grain boundaries in a polycrystalline material.

c) Domain Magnetization

Now when the magnetic field is applied, then the magnetization occurs in the specimen by two ways

1) By moment of domain walls
2) By rotation of domain walls
3) *By moment of domain walls*

The moment of domain walls takes place in weak magnetic fields. Due to this weak field applied to the specimen the magnetic moment increases and hence the boundary of domains displaced, so that the volume of the domains changes.

By Rotation of Domain Walls

The rotation of domain walls takes place in strong magnetic fields. When the external field is high then the magnetization changes by means of rotation of the direction of magnetization towards the direction of the applied field.

d) Energies Involved In Domain Growth

To study the domain structure clearly, we must know four types of energy involved in the process of domain growth. They are:

1) Exchange energy
2) Anisotropy energy
3) Domain wall energy
4) Magneto-strictive energy.

Exchange Energy (or) Magnetic Field Energy (or) Magneto-Static Energy

The interacting energy which makes the adjacent dipoles to align themselves is known exchange energy (or) magnetic field energy. The exchange energy has established a single domain in a specimen of ferromagnetic and it is shown in figure 3.3.4. It is the energy required in assembling the atomic magnets in a single domain and this work done is stored as potential energy.

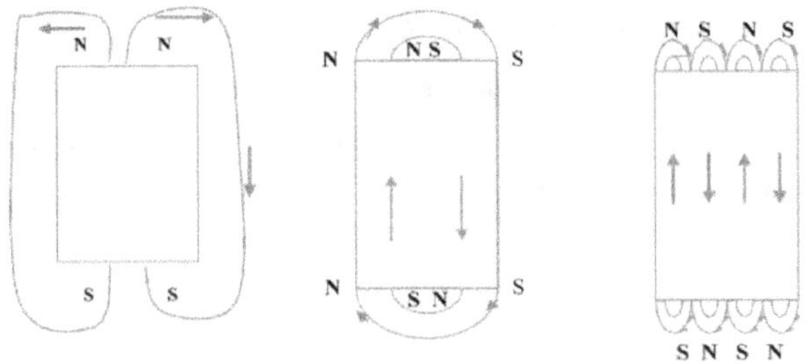

Figure 3.3.4: Representaion of Exchange Energy

Anisotropy Energy

In ferromagnetic crystals there are two direction of magnetization.

1) Easy direction,

2) Hard direction

In easy direction of magnetization, weak field can be applied and in hard direction of magnetization, strong field should be applied. For producing the same saturation magnetization along both hard and easy direction, strong fields are required in the hard direction than the easy direction. For example in iron easy direction is [100], medium direction is [110] and the hard direction is [111] and it is shown in fig. From the figure 3.3.5, we can see that very strong field is required to produce magnetic saturation in hard direction [111] compared to the easy direction [100].

Therefore the excess of energy required to magnetize the specimen along hard direction over that required to magnetize the specimen along easy direction is called crystalline anisotropy energy.

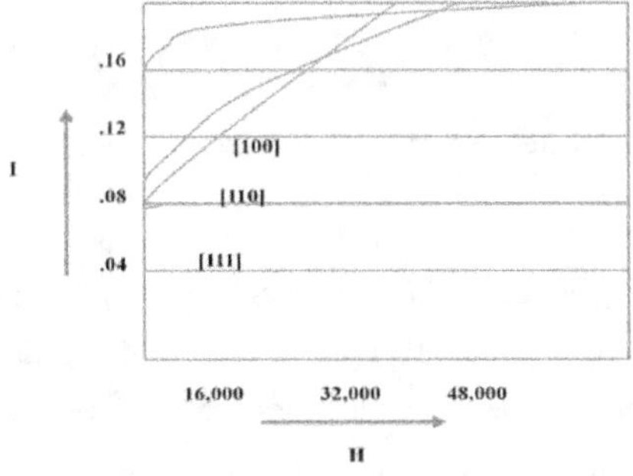

Figure 3.3.5: Representation of Anisotropy Energy

Domain Wall Energy (or) Bloch Wall Energy

Domain wall is a transition layer which separates the adjacent domains, magnetized in different directions. The energy of domain wall is due to both exchange energy and anisotropy energy. Based on the spin alignment, two types of domain walls may arise, namely Thick wall, Thin wall.

a) Thick Wall

When the spin at the boundary are misaligned if the direction of the spin changes gradually as shown in figure 3.3.6, it leads to a thick domain wall. Here the misalignments of spins are associated with exchange energy.

b) Thin Wall

When the spin at the boundaries changes abruptly, then the anisotropy energy becomes very less. Since the anisotropy energy is directly proportional to the thickness if the wall, this leads to a thin Bloch wall.

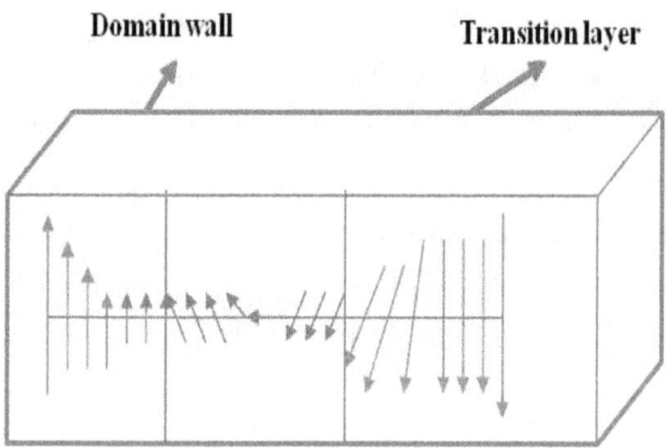

Figure 3.3.6: Representation of Anisotropy Energy

Magnetostrictive Energy

When the domains are magnetized in different directions, they will either expand (or) shrink. Therefore there exits a deformations (i.e) change in dimension of the material, when it is magnetized. This phenomenon is known as magnetosriction and the energy produced in this effect is known as magnetostriction energy. The deformation is different along different crystal directions and the change dimension depends upon the nature of the material.

e) *Explanation of Hysteresis Based on Domain Theory*

Hysteresis

When a ferromagnetic material is made to undergo through a cycle of magnetization, the variation of magnetic induction (B) with respect to applied field (H) can be represented by a closed hysteresis loop (or) curve. (i.e) it refers to the lagging of magnetization behind the magnetizing field. If magnetizing field (H) is applied to a ferromagnetic material and if H is increases to H_{max} the material acquires magnetism. So the magnetic induction also increases, represented by oa in the figure 3.6.

Now if the magnetic field is decreased from Hmax to zero, the magnetic induction will not fall rabidly to zero, but falls to 'b' rather than zero. This shows that even when the applied field is zero or removed, the material still acquire some magnetic induction (ob) which is so called residual magnetism or retntivity. Now, to remove the residual magnetism, the magnetic field strength is reversed and increased to $-H_{max}$ represented as 'oc' so called coercivity we get the curve 'bcd'. Then the reverse field (-H) is reduced to zero and the corresponding curve 'de' is obtained and by further increasing H to Hmax the curve 'efa' is obtained. We know when the ferromagnetic material is subjected to external field, there is an increase in the value of the magnetic moment due to two process.

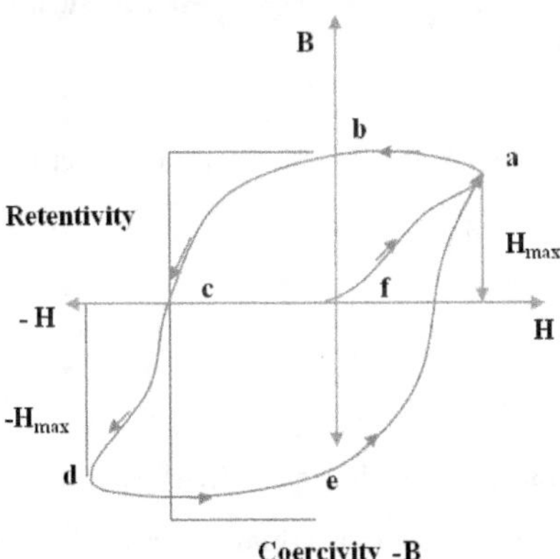

Figure 3.3.7: Representation of Hysteresis Nature of Material

(i) The moment of domain walls (ii) Rotation of domain walls. When small external field is applied, the domains walls displaced slightly in the easy direction of magnetization. This gives rise to small magnetization corresponding to the initial portion of the hysteresis curve (OA) as shown in figure 3.3.7. Now of the field is removed, then the domains returns to the original state, and is known as reversible domains. When the field is increased, large numbers of domains contribute to the magnetization and thus the magnetization increases rabidly with H. Now, even when the field is removed, because of the displacement of domain wall to a very large distance, the domain boundaries do not come back to their original position. This process is indicating as AB in fig and these domains are called irreversible domains.

Now, when the field is further increased, the domains starts rotating along the field direction and the anisotropic energy stored in the hard direction, represented as BC in the figure 3.8. Thus the specimen is said to attain the maximum magnetization. At this position, even when the field is removed the material posses maximum magnetization, called residual magnetism or retntivity, represented by OD in figure 3.8. Actually after the removal of external field, the specimen will try to attain the original configuration by the moment of domain wall. But this moment is stopped due to presence of impurities, lattice imperfections etc. therefore to overcome this; a large amount of reverse magnetic field is applied to the specimen. The amount of energy spend to reduce the magnetization to zero is called coercivity represented by OE in the figure 3.3.8.

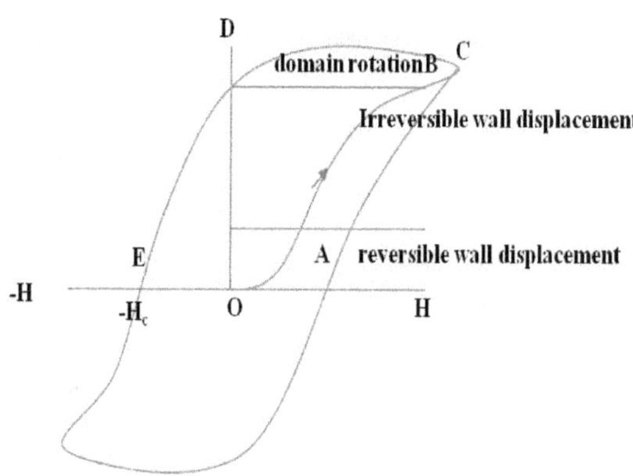

Figure 3.3.8: Representation of Hysteresis Nature of Material

It is the loss of the energy in taking a ferromagnetic specimen through a complete cycle of magnetization and the area enclosed is called hysteresis loop.

7) *Soft and Hard Magnetic Materials*

Depending upon the direction of magnetization by external field, and the area of hysteresis, magnetic can be classified into two types as,

S.No	Soft magnetic materials	Hard magnetic materials
1.	The magnetic materials can be easily magnetize and demagnetize. Example – pure iron, cast iron, carbon steel, silicon steel, mumetal.	The magnetic materials cannot be easily magnetize and demagnetize. Example – tungsten steel, cobalt steel, alini, alnico, hypernic.
2.	They have high permeability.	They have low permeability.
3.	Magnetic energy stored is not high.	Magnetic energy stored is high.
4.	Low hysteresis loss due to small hysteresis loop area.	Large hysteresis loss due to large hysteresis loop area.
5.	Coercivity and retentivity are small.	Coercivity and retentivity are large.
6.	The eddy current loss is small due to its high resistivity.	The eddy current loss is high due to its low resistivity.
7.	The domain walls are easy to move.	The movement of domain wall must be prevented.
8.	They are used in electric motor, generators, transformers, relays, telephone receivers, radar.	They are used in loud speakers and electrical measuring instruments.

a) Energy Product

Definition

The product of residual magnetic induction (B_r) and coercivity (H_c) is called energy product or BH product. It is the important quantity to design powerful permanent magnets. It gives the maximum amount of energy stored in the specimen.

Explanation

The energy required to demagnetize a permanent magnet is given by the area of the hysteresis loop between B_r and H_c. The maximum value of this area $B_r H_c$ is called the energy product. At C and D the energy product is zero because at C, H value is zero and D, B value is zero. The area occupied by the largest rectangle in demagnetizing curve gives the maximum (BH) value. The energy product is large for permanent magnets. This value is very much useful to analyze whether the material can be used for magnetic recording.

8) Ferrimagnetic Materials-Ferrites

a) Ferrites

Ferrites are components of iron oxide with oxides of other components. The general chemical formula is $X^{2+} Fe_2^{3+} O_4^{2-}$, where ($X^{2+}$) is a divalent metal ion such as Fe^{2+}, Mg^{2+}, Ni^{2+}, Co^{2+}, Mn^{2+}.

Structure of Ferrites

Generally there are two types of structures present in the ferrites. They are (i) Regular spinal, (ii) Inverse spinal

Regular Spinal

In regular spinal structure, each divalent metal ion is surrounded by four O^{2-} ions in a tetrahedral fashion. For example in Mg^{2+} Fe_2^{3+} O_4^{2-}, the structure of Mg is given in the figure 3.9 (a) and it is called A site. Each Fe^{3+} (trivalent metal ion) is surrounded by six O^{2-} ions and forms an octahedral fashion as shown in figure 3.9 (a). Totally there will be 16 such octahedral sites in the unit cell. This is indicated by B site. Thus in regular spinal, each divalent metal ion (Mg^{2+}) exits in tetrahedral form (A site) and each trivalent metal ion (Fe^{3+}) exits in an octahedral form (B site). Hence the sites A and B combine together to form a regular spinal ferrite structure as shown in figure 3.9 (b).

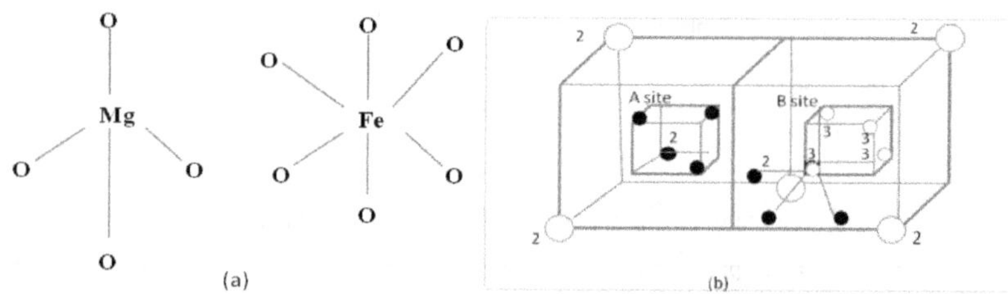

Figure 3.3.9: Representation of Regular Spinal Structure

Inverse Spinal

In this type, we consider the arrangement of ions of a single ferrous ferrite molecule Fe^{3+} [Fe^{2+} Fe^{3+}] O_4. A Fe^{3+} ion (trivalent) occupies all A sites (tetrahedral) and half of the B sites (octahedral) also. Thus the left out B sites will be occupied by the divalent (Fe^{2+}). The inverse spinal structure is shown in figure 3.3.9.1.

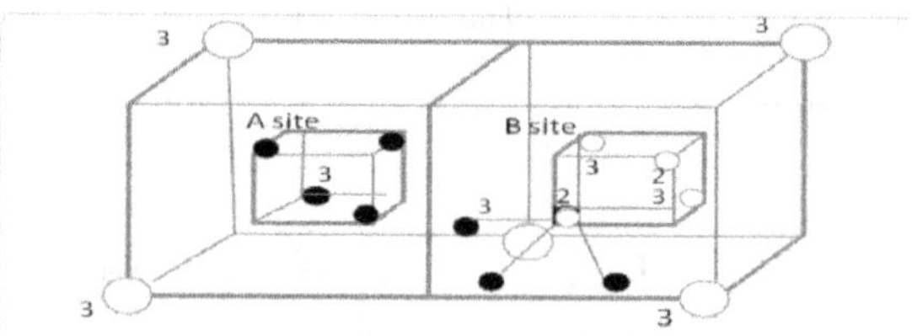

Figure 3.3.9.1: Representation of Inverse Spinal Structure

Preparation

They have the general chemical composition $A^{2+} Fe_2^{3+} O^{2-}$ where A^{2+} represent a divalent metal ion like Zn^{2+}, Mg^{2+}, etc. Ferrities are prepared by sintering a mixture of various metallic oxides as follows.

1) Suitable of A^{2+} and $Fe_2^{3+} O_4$ in proper proportions are mixed using water or kerosene.
2) The mixing is done in a blender for several hours. It is filtered.
3) The filtered material is dried in a hot oven and is then crushed.
4) Next mixture, is pre-sintered in a furnace at $900^0 c$ to $1100^0 c$ for a period of three to fifteen hours, in an air atmosphere or nitrogen atmosphere.
5) The pre-sintered material is then ground into a fine powder and mixed with a binder such as paraffin wax and solvent such as water.
6) The mixture is then pressed into the desired shapes by using dies.
7) The last step in the process is to place the ferrite in proper vessel in a furnace and heat it to about 1100 to $1400^0 c$. The binder then evaporates. It is cooled in a controlled manner.

Properties of Ferrites

1) Ferrites have net magnetic moment.
2) Above Curie temperature, it becomes paramagnetic, while it behaves ferromagnetic material blow Curie temperature.
3) The susceptibility of ferrite is very large and positive. It is temperature dependent and is given by

$$\chi = C / T \pm \theta \qquad \text{for } T > T_N$$

4) Spin alignment is antiparallel of different magnitudes as shown below.
5) Mechanically it has pure iron character. They have high permeability and resistivity.

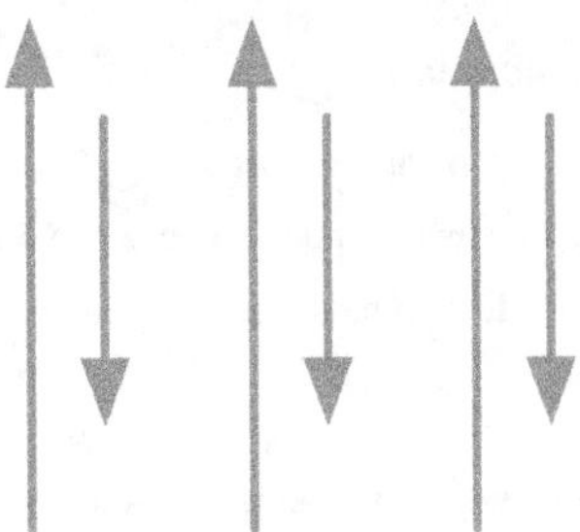

6) They have low eddy current losses and low hysteresis losses.

Advantages

1) Efficiency is high and cost is low.
2) They have low eddy current losses and low hysteresis losses.
3) Easy to manufacture with great uniformity.
4) They occupy low volume.

3.4. Super Conducting Materials

The electrical resistivity of metals increases linearly with increasing temperature near room temperature whereas the resistivity of semiconductors decreases non linearly with increasing temperature. For certain materials, like mercury, the resistivity suddenly drops to zero at very low temperature typically near the boiling point of liquid helium. This phenomenon is known as superconductivity. Some metals, doped semiconductors, alloys and ceramics show superconductivity. Super conductivity is a phenomenon which is much different from conductivity as some of the best conductors like gold, silver and copper do not show superconductivity at any temperature whereas some ceramics which are insulators at room temperature show superconductivity. Infact, the ceramics become superconducting at higher temperature than metals.

Super conducitvity has vast potential importance in technology as current can flow through a superconductor without any thermal losses. However, applications on a large scale have not been possible due to requirement of very low temperatures.

Super conductivity is one of the most existing phenomena in physics. It was discovered by Dutch physicist H.K Onnes in the year 1905.

Before the discovery of super conductivity, it was though that the electrical resistance is zero only at absolute zero. But, it was found that in some material the electrical resistance becomes zero, when they are cooled to very low temperature.

1) Superconducting Phenomena

Super conducting materials have extraordinary electrical and magnetic characteristics. These materials have many important applications in the field of engineering and technology. Many electronic and magnetic devices have been fabricated with super conducting materials.

2) Definition

The phenomenon of sudden disappearance of electrical resistance in a material, when it is cooled blow a certain temperature is known as super conductivity.

3) *Transition Temperature or Critical Temperature*

The temperature at which a material at normal conducting state changes into superconducting state is known as transition temperature or critical temperature (T_c).

The transition temperature depends on the property of the material. It is found that the super conducting transition is reversible, i.e., above critical temperature (T_c) a super conductor becomes a normal conductor.

It is to note that the metals which are normally very good conductors of heat and electricity (e.g. Cu. Ag, Au) are not super conductors.

4) *General Features of Superconductors*

1) Superconductivity is found to occur in metallic elements in which the number of valence electron lies between 2 and 4.
2) Materials having high normal resistivities exhibit superconductivity.
3) Superconductivity is also favoured by small atomic volume accompanied by a small atomic mass.
4) The transition temperature (T_C) is different for different substances.
5) Ferromagnetic and anti ferromagnetic materials are not superconductors.
6) The electrical resistivity drops to zero.
7) The magnetic flux lines are expelled from the material.
8) There is a discontinuous change in the specific heat.
9) Further, there are some small changes in the thermal conductivity and the volume of the materials.

3.4.1. *Properties of Super Conductors*

(i) *Zero Electrical Resistance*

The first characteristic property of a super conductor is its electrical resistance. The electrical resistance of the super conductor is zero below a transition temperature. This property of zero electrical resistance is known as defining property of a superconductor.

The variation of electrical resistivity of a normal conducting metal and a superconducting metal as a function of temperature is shown in fig

(ii) *Effect of Magnetic Field*

Below the transition temperature of a material, its superconductivity can be destroyed by the application of a strong magnetic field. The minimum magnetic field strength required to destroy the superconducting property at any temperature is known as critical magnetic field (Hc).

The critical magnetic field (Hc) depends upon the temperature of the superconducting material. The relation between critical magnetic field and temperature is given by

$$H_c = H_o \left[1 - (T / T_c)^2\right]$$

Where, H_o- is critical magnetic field at absolute zero temperature. T_c- is superconducting transition temperature of a material

T – is the temperature below T_c of the superconducting material.

(iii) Effect Electric Current

The application of very high electrical current to superconducting material destroys its superconducting property.

The critical current (I_c) is required to destroy the superconducting property is given by

$$I_c = 2\pi r H_c \quad \text{Where } H_c - \text{ is the critical magnetic field}$$
$$r \text{ - is the radius of the superconducting wire}$$

(iv) Persistent Current

A steady current which flows through a superconducting ring without any decrease in its strength as long as the material is in superconducting state is called persistent current. The current persists even after the removal of the magnetic field.

(v) Meissner Effect

When a superconducting material is placed in a magnetic field of flux density 'B' the magnetic lines of force penetrates through the material as shown in figure 3.4.1.

Now, when the material is cooled below its transition temperature then the magnetic lines of force are ejected out from the material as shown in fig.

We know that diamagnetic material have the tendency to expel the magnetic lines of force. Since the super conductor also expels the magnetic lines of forces it behaves as a perfect diamagnet. This behaviour is first observed by Meissner and hence called as Meissner effect.

When a superconducting material is placed in a magnetic field, under the condition $T \leq T_c$ and $H \leq H_c$ the flux lines are excluded from the material. Thus the material exhibits perfect diamagnetism. This phenomenon is called as Meissner effect.

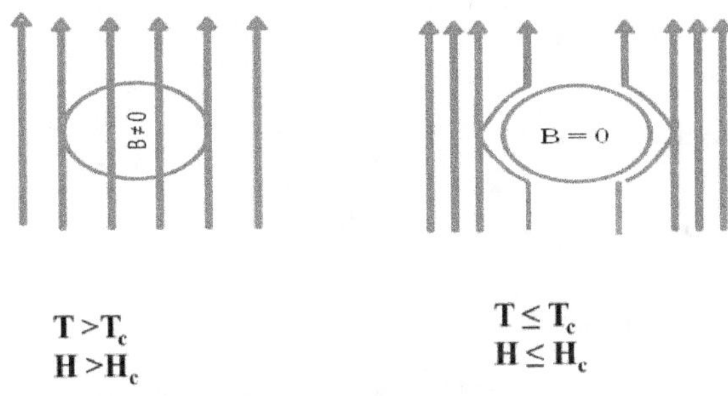

Figure 3.4.1: Representation of Meissner Effect

Proof

We know $B = \mu_o(H + I)$

When $B = 0$ we get $0 = \mu_o(H + I)$

Since $\mu_o \neq 0$ we can write $H + I = 0$

$$\text{Or} \quad -H = I \text{ or } I / H = -1 = \chi$$

Since the susceptibility is negative, this shows that superconductor is perfect diamagnet.

(vi) Isotope Effect

The transition temperature varies due to presence of isotopes.

Example

The atomic mass of mercury varies from 199.5 to 203.4, and hence the transition temperature varies from 4.185 K to 4.146 K. Due to the relationship $T_c \, \alpha \, [1 / M\alpha]$ M–atomic weight α-constant (=.5)

3.4.2. Type-I and Type-Ii Superconductors

There are two types of super conductors based on their variation in magnetization, due to external magnetic field applied.

1) Type I superconductor or soft super conductor.
2) Type II superconductor or hard superconductor.

Type I Superconductor

When the super conductor is kept in the magnetic field and if the field is increased the superconductor becomes normal conductor abruptly at critical magnetic field as shown in figure 3.4.2. These types of materials are termed as Type I superconductors.

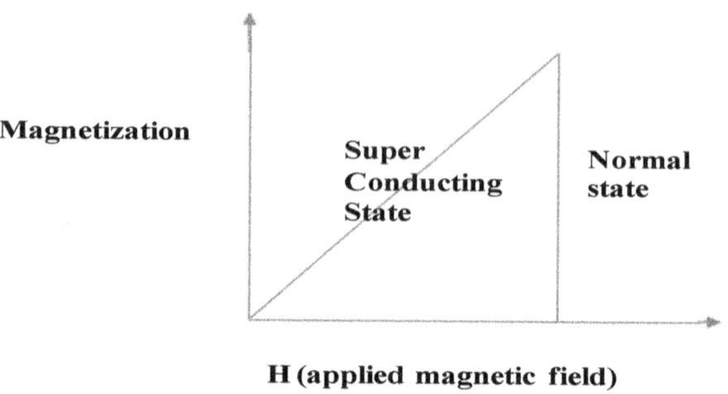

Figure 3.4.2: Representation of Type I Superconductors

Below critical field, the specimen excludes all the magnetic lines of force and exhibit perfect Meissner effect. Hence, Type I superconductors are perfect diamagnet, represented by negative sign in magnetization.

Type II Superconductors

When the super conductor kept in the magnetic field and if the field is increased, below the lower critical field Hc1, the material exhibit perfect diamagnetism (i.e) it behaves as a super conductor and above Hc1, the magnetization decreases and hence the magnetic flux starts penetrating through the material. The specimen is said to be in a mixed state between Hc1 and Hc2. above Hc2 (upper critical field) it becomes normal conductor as shown in figure 3.4.3.

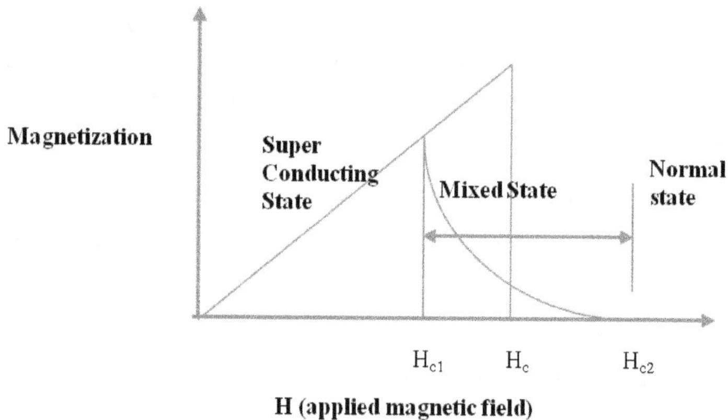

Figure 3.4.3: Representation of Type II Superconductors

The materials which lose its superconducting property gradually due to increase on the magnetic field are called Type II superconductor.

3.4.3. Difference between Type–I and Type–II Superconductors

S.No	Type – I Superconductors	Type – II Superconductors
1.	The material loses magnetization suddenly.	The material loses magnetization gradually.
2.	They exhibit complete Meissner effect i.e., they are completely diamagnetic.	They do not exhibit complete Meissner effect.
3.	There is only one critical magnetic field (H_c).	There are two critical magnetic fields i.e., lower critical field (H_{c1}) and upper critical field (H_{c2}).
4.	No mixed state exists.	Mixed state is present.

3.4.4. BCS Theory

BCS theory (qualitative)

The microscopic theory of superconductivity developed by J. Bardeen, L.N. Cooper and J.R. Scriffer in 1957, successfully explained the effect like zero resistivity, Meissner effect etc. this theory is known as BCS theory.

Principle

This theory states that the electrons experience a special kind of attractive interaction, overcoming the coulomb forces of repulsion between them; as a result cooper pairs (i.e) electro pair are formed. At low temperature, these pairs move without any restriction through the lattice points and the material becomes superconductor. Here the electron-lattice-electrons interaction should be stronger than electrons-electros interaction.

Important Features of BCS Theory

1) Electrons form pairs (called cooper pair) which propagate throughout the lattice.
2) The propagation of cooper pairs is without resistance because the electrons move in resonance with phonons.

Electron-Lattice-Electron Interaction

When an electron (1st) moves through the lattice, it will be attracted by the core (+ve charge) of the lattice. Due to this attraction, ion core is disturbed and it is called as lattice distortion. The lattice vibrations are quantized in terms of phonons.

The deformation produces a region of increased positive charge. Thus if another electron (2nd) moves through this region as shown in fig. It will be attracted by the greater concentration of positive charge and hence the energy of the 2nd electron is lowered.

Hence two electrons interact through the lattice or the phonons field resulting in lowering the energy of electrons. This lowering of energy implies that the force between the two electrons is attractive. This type of interaction is called electrons-lattice electron interaction. The interaction is strong only when the two electrons have equal and opposite momenta and spins.

Explanation

Consider the 1st electron with wave vector k distorts the lattice, here by emitting phonons of wave vector q. This results in the wave vector k-q for the 1st electron. now if the 2nd electron with wave vector k', seeks the lattice it takes up the energy from the lattice and its wave vector changes k'+q as shown in fig. two electrons with wave vectors k-q and k'+q form a pair known as cooper pair.

Cooper Pair

The pair of electrons formed due to electron-lattice (phonons)-electron interaction (force of attraction) by overcoming the electron-electron interaction (force of repulsion), with equal and opposite momentum and spins (i.e) with wave vector k-q and k'+q, are called cooper pair.

Coherence Length

In the electron-lattice-electron interaction, the electrons will not be fixed, they move in opposite directions and their co-relations may persist over lengths of maximum 10^{-6}m. This length is called coherence length.

Note: BCS theory hold good only for low temperature superconductivity.

3.4.5. High Temperature Superconductor

In a superconductor if the transition temperature is high ie., greater than 20K, then it is called as high temperature superconductor. Earlier it was believed that the superconductivity was only in metals. Surprisingly in 1986, Muller and Bednorz discovered high temperature superconductor in ceramics.They made a particular type of ceramic material from a compound of barium, lanthanum, copper and oxygen(Ba-La-Cu-O). This compound superconductor showed superconductivity even at a temperature as high as 30K. The oxide Y Ba_2 Cu_3 O7 with a T_c of 90K was the most extensively studied high temperature superconductor.

Characteristics of High Temperature Superconductor

1) They have high transition temperature.
2) They have a modified perovskite crystal structure.

3) Formation of the superconducting state is direction dependent.

4) They are oxides of copper in combination with other elements.

5) They are reactive, brittle, and cannot be easily modified or joined.

Preparation of High Tc Ceramic Superconductor $Y\,Ba_2\,Cu_3\,O_{9-x}$

The oxide ($Y\,Ba_2\,Cu_3\,O_{9-x}$) is prepared from compacted powder mixture of Y_2O_3, $BaCO_3$ and CuO in the right proportion and heating them in temperature between 900°C and 1100°C. $BaCO_3$ decomposes at this temperature to BaO and CO_2. This is followed by another annealing treatment at 800°C in an atmosphere of oxygen.

Crystal structure of $Y\,Ba_2\,Cu_3\,O_{9-x}$

Here, the primitive cell is developed by three body centered cubic unit cells stacked one above the other to form a tetragonal (a=b≠c) perovskite structure tripled along the C axis as shown in figure 3.4.4.

Yttrium Atoms

Each yttrium atom is shared by one unit cell. 1 / 1th of the atom is shared by that unit cell. Number of yttrium atoms per unit cell   = 1 / 1 X total number of yttrium atoms

$$= 1 / 1 X 1 = 1$$

Barium Atoms

Each barium atom is one unit cell

1 / 1th of the atom is shared by that unit cell.

Number barium atoms per unit cell = 1 / 1 X total number of yttrium atoms = 1 / 1 X 2 = 2 atoms per unit cell.

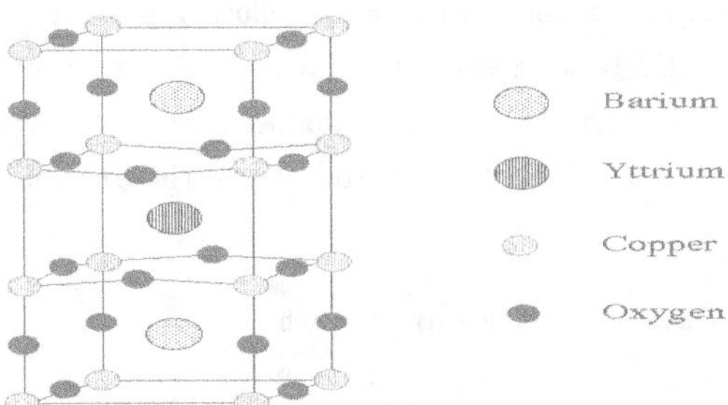

Figure 3.4.4: Representation of Crystal Structure of High Temperature Superconductor

Copper Atoms

Each yttrium atom is shared by 8 unit cells [since copper is the corner atoms].

1 / 8 th of the atom is shared by one unit cell.

Number copper atoms per unit cell = 1 / 8 X total number of yttrium atoms X number of unit cells.

$$= 1 / 8 \text{ X } 8 \text{ X } 3 = 3 \text{ atoms per unit cell.}$$

Oxygen Atoms

Each yttrium atom is shared by 4unit cells [since oxygen atoms are situated at mid points between two corner atoms].

1 / 4th of the atom is shared by one unit cell.

Number copper atoms per unit cell = 1 / 4 X total number of yttrium atoms X number of unit cells.

$$= 1 / 4 \text{ X } 12 \text{ X } 3 = 9 \text{ atoms per unit cell.}$$

3.4.6. Applications of Superconductors

a) Electric Generators

Superconducting generators are very small in size and light weight when compared with conventional generators. The low loss superconducting coil is rotated in extremely strong magnetic field. Motors with very high powers as large as 2500 kw could be constructed at very low as 450 V. This is the basis of new generation of energy saving power systems.

b) Low Loss Transmission Line and Transformers

Since, the resistance is almost zero at superconducting phase, the power loss during transmission is negligible. Hence, electric cables are designed with superconducting wires. If superconductors are used for winding the transformers, the power loss will be very small. Using superconductor, 2000-3000 MW portable transformers have been manufactured.

c) Cryotron

Cryotron is a magnetically operated current switch.

Principle

We know that the superconducting property of a material disappears when the applied magnetic field is greater than the critical magnetic field. Consider a superconducting material A surrounded by another superconducting material B as shown in figure 3.4.5. let the critical magnetic field of material A be less than the critical magnetic field of material B. Initially, let the temperature of the whole system blow the transition temperature of two materials.

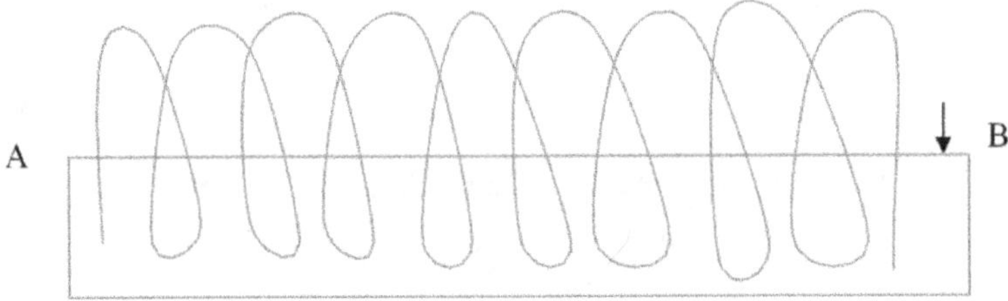

Figure 3.4.5: Representation of Cryotron

Now at the operating temperature, the magnetic field produced by material B may exceed the critical magnetic field of material A. Hence, material A becomes a normal conductor because the critical magnetic field of A is less than that of B. Moreover, B not become a normal conductor at the critical magnetic field of A because HcB > HcA. Therefore the current in the material A can be controlled by the current in the material B. hence this system can act as a relay or switching element.

d) Josephson Devices

Principle

A steady and undiminishing current (Persistent of current) influenced by d.c. voltage is the principle used in Josephson devices.

Josephson Effect

Josephson effect happens by virtue of quantum tunnelling of Cooper pairs. According to Josephson effect, the tunnelling cooper pairs would take place between two superconductors separated by an insulator even in the absence of applied voltage. Pairs of electrons move through the potential barrier induce the superconducting current. This effect is know as Josephson effect.

Explanation

It consists of a thin layer of oxide material placed in between tow superconducting materials as shown in. Here, the Oxide layer acts as a barrier to the flow of conduction electrons from one superconductor to the other.

When the battery is switched ON, the voltage V is applied across the super conductors. Due to applied voltage, the electrons in the super conductor–1 are tunnel across the insulator into the super conductor–2. This tunneling effect produces the current between the superconductors. The increase in voltage produces more and more electrons and hence increases the current. This current has two components.

e) SQUID

SQUID stands for Superconducting Quantum Interference Device. It is an ultra-sensitive instrument used to measure very weak magnetic field in the order of 10^{-14} tesla.

Principle

We know that small change in magnetic field produces in the flux quantum

Description and Working

A SQUID consist of superconducting wire which can have the magnetic field of quantum values (1, 2, 3...) of flux placed in between two Josephson junctions as shown in fig. Then magnetic field if applied perpendicular to the plane of the ring, the current is produced at the two Josephson junctions. The induced current produces the interference pattern and it flows through the ring so that the magnetic flux in the ring can have the quantum value of magnetic field applied.

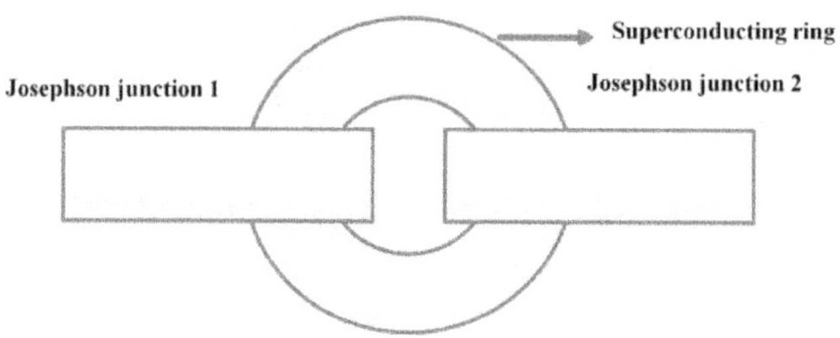

Figure 3.4.6: Representation of SQUID

Application

1) It can be used to detect the variation of very minute magnetic signals in terms of quantum flux.
2) It can also be used as storage device for magnetic flux.
3) It is useful in the study of earthquakes, removing paramagnetic impurities, detection of magnetic signals from the brain, heart.

f) *Magnetic Levitated Train (MAG LEV)*

Magnetic levitated train is the train which cannot move over the rail, rather it floats above the rail, under the condition, when it moves faster.

Principle

Electromagnetic induction is used as the principle. (i.e) when there is a relative motion of a conductor across the magnetic field, current is induced in the conductor and vice versa.

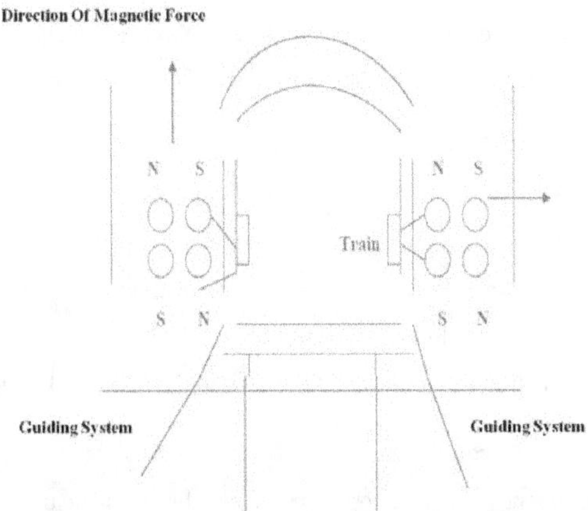

Figure 3.4.7: Representation of MAG-LEV

Explanation

1) This train consists of superconducting magnets placed on each side of the train. The train can run in a guidance system which consists of a series of 8 shaped coils as shown in figure 3.4.7.

2) Initially when the train starts, they slide on the rails. Now, when the train moves faster, the superconducting   magnets on each side of the train will induce a current in the "8" shaped coils kept in the guidance system.

3) This induced current generates a magnetic force in the coils in such a way that the lower half of the 8-shaped coil has the same magnetic pole as that of the superconducting magnet in the train, while the upper half has the opposite magnetic pole.

4) Therefore the total upward magnetic force acts on the train and hence the train is levitated (or) raised above the wheels (i.e) the train now floats above the air.

5) Now, by alternatively changing the poles of the superconducting magnet in the train alternating current can be induced in the "8" shaped coils.

6) Thus, alternating series of north and south magnetic poles are produced in the coils which pulls and pushes the superconducting magnets in the train and hence the train is further moved.

7) The magnetic levitated train can travel a speed 500 km/hour, which is double the speed of existing fastest train

Note: The train is supposed to move always at the centre. Suppose if it moves away from the centre, say for example right side, an attractive force is given at the left side, and a repulsive force is given at the right side and is made to come at the centre.

g) *Engineering Applications*

1) They are used to construct very sensitive electrical measuring instruments such as galvanometers because they require very small voltages.

2) They are used to transmit power over very long distance without any power loss.

3) They are used as storae devices in computers.

4) Superconductors are used to design rectifiers, logic gates, modulators etc.

h) *Medical Applications*

1) They are used to study tiny magnetic signals from brain and heart.

2) They are used in NMR imaging systems.

3) They are used to detect brain tumours and clots using superconducting solenoids.

4) They are used in magneto-cardiography, magneto–encephalography.

5) They are used to study the amount of iron held in the lever of the body accurately.

UNIT-IV

DIELECTRIC MATERIALS

Abstract

After studying this unit, one can acquire

1) Understand about dia, para and ferro magnetism.
2) Acquire knowledge on applications of ferrites.
3) Acquire proper knowledge on dielectrics and their properties.
4) Understand the concepts of dielectric breakdown.

4.1. Introduction

Solids which have an energy gap of three eV or more are termed as insulators or dielectrics. They are capable of storing electrical energy. The dielectric materials do not have free electrons and hence are called good insulators. When they are placed in electric fields, internal fields are set up in the dielectric materials which oppose the externally applied field, thereby reducing the net electric field and hence the potential difference.

If these dielectrics are placed between plates of a capacitor, the potential difference will be reduced without affecting the charge on the plates. Thus, the capacitance of the capacitor increases.

Properties

1) Generally, the dielectrics are non-metallic materials of high resistivity.
2) They have a very large energy gap (more than 3eV).
3) All the electrons in the dielectrics are tightly bound to their parent nucleus.
4) As there are no free electrons to carry the current, the electrical conductivity of dielectrics is very low.
5) They have negative temperature coefficient of resistance and high insulation resistance.

4.1.1. Basic Definitions

i. Electric Dipole

A system consisting of two equal and opposite charges n (+q, -q) separated by a distance (d) is called an electric dipole as shown in figure 4.1.

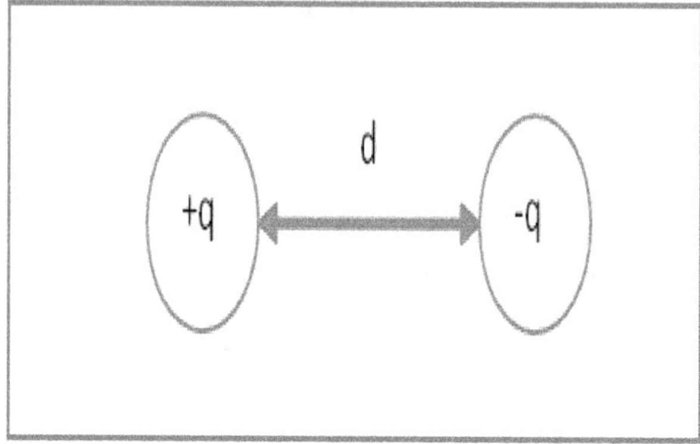

Figure 4.1: Representation of an Electric Dipole

ii.   *Dipole Moment (M)*

The product of the magnitude of the charge (q) and distance between two charges (d) is called as dipole moment.

$$\text{Dipole } moment = qd \ (\text{coulomb-metre})$$

iii.   *Permittivity (ε)*

The permittivity represents the dielectric property of a medium. It indicates easily polarisable nature of material.

Its unit is farad/metre.

iv.   *Dielectric constant (εr)*

1) A dielectric characteristic of a material is determined by its dielectric constant. It is a measure of polarisation of the dielectrics.
2) It is the ratio between absolute permittivity of the medium (ε) and permittivity of free space (ε_o).

$$\text{Dielectric constant} = \frac{\text{Absolute Permitivitty } (\varepsilon)}{\text{Permittivity of free space } (\varepsilon o)}$$

$$\varepsilon_r = \varepsilon / \varepsilon_o$$

v.   *Polarization*

The process of producing electric dipoles inside the dielectric by the application of an external electrical field is called polarization in dielectrics.

vi. *Polarisability (α)*

It is found that the average dipole moment (μ) of a system is proportional to the applied electric field (E).

$$\mu \, \alpha \, E.$$

$$\text{or} \quad \mu = \alpha \, E$$

Where (α) is the polarisability.

$$\alpha = \mu \, / \, E$$

Polarisability is defined as the ratio of average dipole moment to the electrical field applied. Its unit is farad m^2

vii. *Polarisation vector [P]*

It is defined as the average dipole moment per unit volume of a dielectric. If N is the number atoms per unit volume of a dielectric and (μ) is average dipole moment per atom, then

$$\vec{P} = N\mu \, (\text{coulomb} \, / m^2)$$

4.2. Various Polarization Mechanisms Involved in Dielectric

Dielectric polarization is the displacement of charged particles under the action of the external electric field. There are number of devices based on this concept. Those devices are rectifiers, resonators, amplifiers and transducers, which converts electrical energy in to other forms of energy. Dielectric polarization occurs due to several microscopic mechanisms.

1) Electronic polarization
2) Ionic polarization
3) Orientational polarization
4) Space-charge polarization

1) Electronic Polarization

Electronic polarization occurs due to displacement of positively charged nucleus and negatively charged electrons of an atom in the opposite directions on the application of an electric field. This will result in the creation of dipole moment in the dielectric.

Dipole moment (μ) is proportional to the electric field strength (E).

$$\mu \, \alpha \, E$$

$$\mu = \alpha_e \, E$$

Where (α_e) is proportionality constant and it is known as electronic polarizability. Electronic polarization takes place in almost all dielectrics.

Expression for Electronic Polarizability

Without Electric Field

Consider an atom of a dielectric material of nuclear charge Ze, where Z is the atomic number. The electrons of charge (-Ze) are distributed uniformly throughout the atom (sphere) of radius R as shown in figure 4.2.

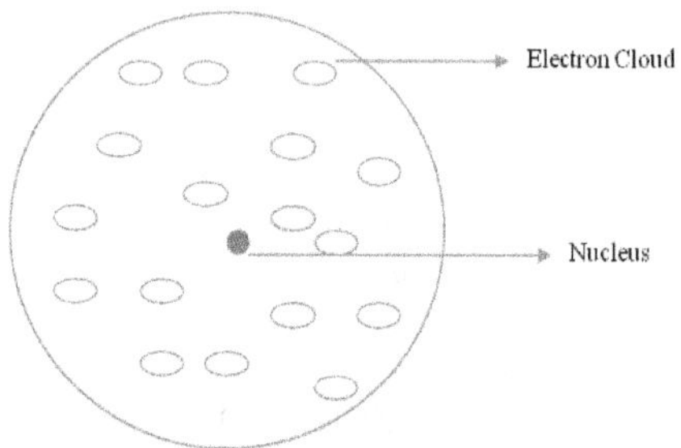

Figure 4.2: Representation of An Atom

The centres of the electron cloud and the positive nucleus are at the same point and hence there is no dipole moment.

Negative charge density of an atom of radius R is given by,

$$\rho = \frac{\text{Total negative charge}}{\text{Volume of the atom}} = \frac{-Ze}{\frac{4}{3}\pi R^3}$$

$$\rho = \frac{4}{3}\left(\frac{-Ze}{\pi R^3}\right) \qquad (1)$$

With Electric Field

When the atom of the dielectric is placed in an electric field of strength E, two phenomenona occur.

Lorentz force (due to electric field) will tend to move the nucleus and electron cloud of that atom from their equilibrium positions. The positive nucleus will move towards the field direction and the electron cloud will move in the opposite direction of the field as shown in figure 4.3.

After separation, an attractive coulomb force arises between the nucleus and the electron cloud which will tend to maintain the original equilibrium position.

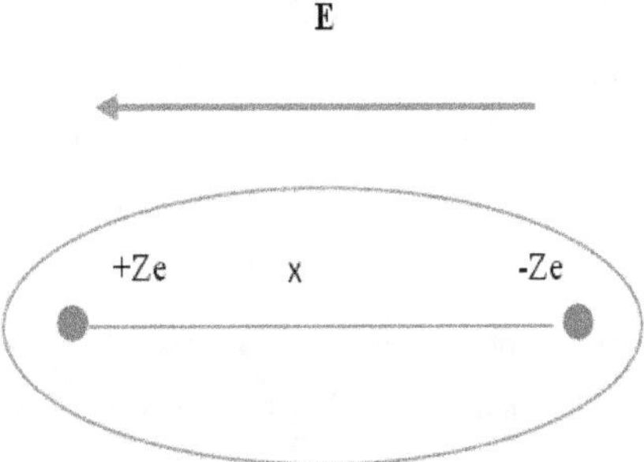

Figure 4.3: Representation of an Atom in the Presence of Electric Field

The electron cloud and the nucleus move in opposite directions and they are separated by a distance x, where there is a formation of electric dipole in the atom.

When these two forces equal and opposite, there will be a new equilibrium between the nucleus and the electron cloud of the atom.

Lorentz force between the nucleus and the electron F_L = charge x electrical field

$$= ZeE \qquad\qquad (2)$$

Coulomb attractive force between the nucleus and the electron cloud being separated at a distance x,

$$F_C = \frac{\text{Charge of the nucleus } \times \text{ total negative charges enclosed in the sphere of radius R}}{(4\pi\varepsilon_o x^2)}$$

$$F_C = \frac{Ze}{4\pi\varepsilon_o x^2} \times \frac{-Zex^3}{R^3} = \frac{-Z^2 e^2 x}{4\pi\varepsilon_o R^3} \qquad (3)$$

At equilibrium, Lorentz Force = Coulomb force

$$-ZeE = \frac{-Z^2 e^2 x}{4\pi\varepsilon_o R^3}$$

$$x = \frac{4\pi\varepsilon_o R^3 E}{Ze}$$

Thus the displacement of the electron cloud is proportional to the applied field.

Therefore, Induced dipole moment $= \mu_e = Ze \cdot x$

$$= \frac{Ze\, 4\pi\varepsilon_o R^3 E}{Ze}$$

$$\mu_e = 4\pi\varepsilon_o R^3 E$$

Since, $\mu_e = \alpha_e E$

Therefore, $\alpha_e = 4\pi\varepsilon_o R^3$

This is the expression for electronic polarization.

Where $\alpha_e = 4\pi\varepsilon_o R^3$ is called electronic polarizability.

1) Electronic polarization is independent of temperature.
2) It is proportional to the volume of atoms in the material. Electronic polarization takes place in all dielectrics.

2) Ionic Polarisation

The process of displacement of positive ions and negative ions in the opposite directions when an external electric field is applied and thereby creates a dipole moment in the dielectric is called ionic polarization.

Induced dipole moment, $\mu = \alpha_i E$

α_i – ionic polarisability

Explanation

1) In the Absence of Electric Field: Consider NACL Molecule

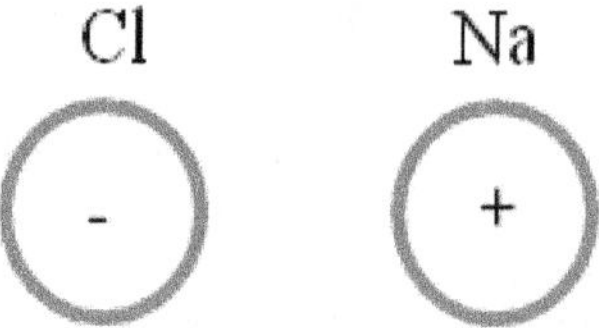

Figure 4.4: Representation of Molecule in the Absence of Electric Field

The ions will be in their original position as shown in figure.

2) In the Presence of Electric Field

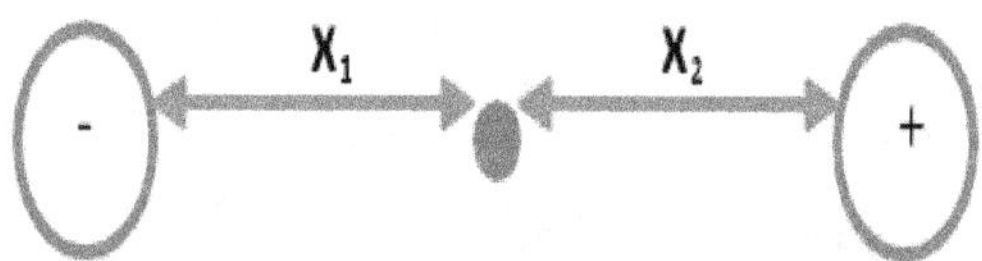

Figure 4.5: Representation of Molecule in the Presence of Electric Field

Let the field is applied along X- direction. The +ve ion move to the right by X_2. The – ve ion move to the left by X_1.

Thus, the net displacement is $X_1 + X_2$.

Since, Displacement α restoring force,

Thus,

For Positive ions	For Negative ions
Restoring force,	Restoring force,
$F \propto X_2$	$F \propto X_1$
$F = \beta_2 X_2$	$F = \beta_1 X_1$
$X_2 = F / \beta_2$	$X_1 = F / \beta_1$
β_1, β_2- Restoring force constant.	
Since, $\beta_1 = m \, \omega_o^2$; $\beta_2 = M \, \omega_o^2$ & $F = eE$.	
Therefore,	
$X_2 = eE / M \, \omega_o^2$ (1)	$X_1 = eE / m \, \omega_o^2$ (2)

Therefore,

$$X_1 + X_2 = eE / m \, \omega_o^2 + eE / M \, \omega_o^2 = eE / \omega_o^2 \, [(1/m) + (1/M)] \qquad (3)$$

Since, The net resultant dipole moment, μ = Charge X Displacement

$$\mu = e \, (X_1 + X_2)$$
$$\mu = e2E / \omega o2 \, [(1/m) + (1/M)]$$
$$\mu = \alpha_i E$$
$$\text{Therefore, } \alpha_i = e2 / \omega o2 \, [(1/m) + (1/M)]$$

This is the expression for ionic polarisability in dielectric.

Conclusion

1) Ionic polarizability (α_i) is inversely proportional to the square of angular frequency of the ionic molecule.
2) It is directly proportional to its reduced mass given by $[(1/m) + (1/M)]$
3) It is independent of temperature.

3) Oreintaional Polarization

Orientational polarization takes place only in polar dielectrics. Polar dielectrics have molecules with permanent dipole moments even in the absence of external electric field.

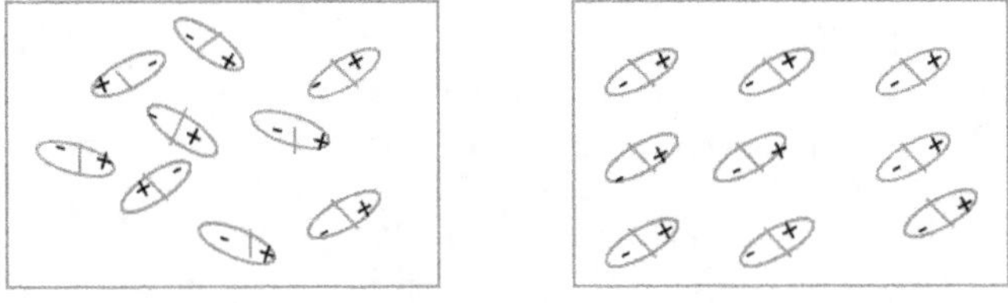

Figure 4.5: Representation of Orientation Polarization in Molecule

When the polar dielectrics are subjected to external electric field, the molecular dipoles are oriented in the direction of electric field as shown in figure 4.5. The contribution to polarization due to orientation of molecular diploes is called orientational polarization.

Orientational polarization depends upon temperature when the temperature is increased , thermal energy tends to disturb the alignment.

From the Langevin's theory of paramagnetism , net intensity of magnetization

$$= N\mu^2 B / 3kT \qquad (1)$$

Since the same principle can be applied to the application of electric field in dielectrics, we may write

Orientational polarization, $P_o = N\mu^2 E / 3kT$

But, Orientational polarization is proportional to applied field (E) and it is given by

$$P_o = N \, \alpha_o \, E \qquad (2)$$

Comparing equations 1 and 2, we get

$$\alpha o \ = \mu 2 / \ 3kT$$

Where (α_o) is orientational polarizability.

Conclusion

The orientation polarizability is inversely proportional to absolute temperature of the material.

4) Space Charge Polarization

Space-charge polarization occurs due to accumulation of charges at the electrodes or at the interfaces of multiphase dielectric material. When such materials subjected to an electrical field at high temperature, the charges get accumulated as shown in figure 4.6. these charges create diploes.

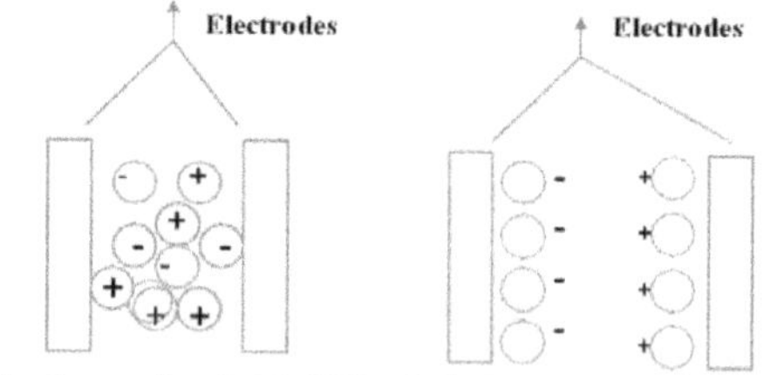

Figure 4.6: Representation of Space Charge Polarization

As a result, polarization is produced. This kind of polarization is known as space-charge polarization. Space-charge polarization is very small when compared to other polarization mechanisms and it is not common in most of the dielectrics. E. g- ferrites and semiconductors.

Total Polarization

$$\alpha = \alpha e + \alpha i + \alpha o$$

Since the space- charge polarization is very small and it is negligible.

Substituting the corresponding expressions, we have

$$\alpha = 4\pi\varepsilon_o R^3 + \mu^2 / 3kT + e^2 / \omega_o^2 [(1/m) + (1/M)]$$

Since, Total Polarisation, $P = N\alpha E$

Therefore, $P = NE (4\pi\varepsilon_o R^3 + \mu2 / 3kT + e2 / \omega o2 [(1/m) + (1/M)])$

This equation is known as Langevin–Debye equation.

4.3.   Active and Passive Dielectrics

The dielectric materials may be classified as solid, liquid and gas dielectrics.

In solid form they may be polymeric such as nylon, pvc, rubber, Bakelite, asbestos and wool or may belong to the ceramic family such as glass, silica, mica, porcelain, etc.

In liquid form they may be mineral insulating oils, synthetic insulating oils, tec.

In gaseous form they may be air, nitrogen, sulphur hexafluoride, inert gases etc.

The dielectrics can also be classified as active and passive dielectrics based on their applications.

1) *Active Dielectrics*

When dielectric is subjected to external electric field, if the dielectric actively accepts the electricity, then they are termed as active dielectrics. Thus active dielectrics are the dielectrics which can easily adapt itself to store the electrical energy in it.

Examples: Piezo-electrics, Ferro-electric etc.

2) *Passive Dielectrics*

These dielectrics are also called insulating materials. As the name itself suggests that it will act as an insulator, conduction will not take place through this dielectrics. Thus passive dielectrics are the dielectrics which restrict the flow of electrical energy in it.

Examples: All insulating materials such as glass, mica, etc.,

4.4. Frequency and Temperature on Polarisaion of Dielectrics

When an alternating electric field is applied across the material, polarization occurs as a function of time. Polarization as a function of time t is given by

$$P(t) = P[1 - e(-t/tr)]$$

Where P is maximum polarization that occurs due to the static field applied for a long time. tr is relaxation time.

1) *Frequency Dependence*

1) Electronic polarization is the fastest polarization which will complete at the instant the field is applied. The reason is that the electrons are lighter elementary particles than ions.

2) Therefore even for very high frequency applied (in the optical range) electronic polarization occurs during every cycle of the applied field.

3) Ionic polarization is little slower than electronic polarization. Because ions are heavier than the electron cloud, the time taken for displacement is large. In addition the frequency of applied field with which the ions will be displaced is equal to the frequency of lattice vibration (10^{13} Hz).

4) If the frequency of the applied field is less than 10^{13} Hz, the ions have enough time to respond during each cycle of the applied field.

5) Orientational polarization is even slower than ionic polarization. This type of polarization occurs only at electrical frequency range ($= 10^6$Hz).

6) Space-charge polarization is the slowest because have to diffuse over several atomic distance.

This process occurs at very low frequencies (10^2 Hz) as shown in figure 4.7.

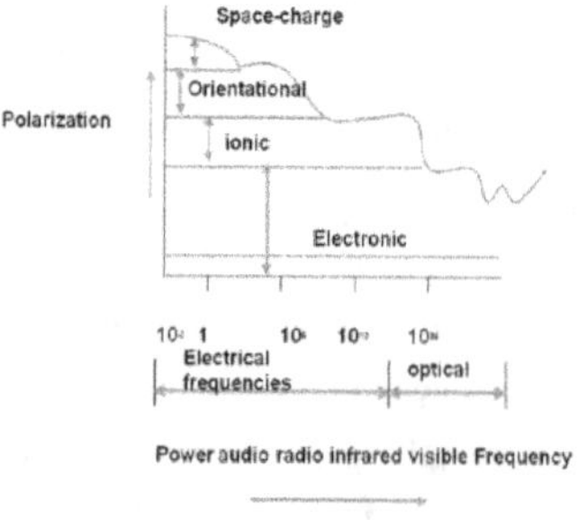

Figure 4.7: Representation on Frequency and Temperature on Polarisaion of Dielectrics

1) Figure 4.7 explains the four types of polarization at different frequency ranges. At optical frequencies (-10^{15} Hz), electronic polarization alone present. At -10^{13} Hz range, ionic polarization occurs in addition to electronic polarization.

2) At 10^6 to 10^{10} Hz range, contribution due to orientation polarization gets added while at 10^2 Hz range, space-charge polarization also contributes.

3) It is noted that at low frequencies, all the four types of polarizations occur and total polarization is very high.

4) Total polarization decreases with increase in frequency and becomes minimum at optical frequency range.

2) Temperature Dependence

1) When a dielectric material is subjected to ordinary conditions of increasing temperature, electronic distribution in n the constituent molecules are not affected.

2) Hence there will be no temperature influence on electronic and ionic polarization mechanisms.

3) Therefore electronic and ionic polarizations are practically independent of temperature.

4) An increase in temperature brings a high degree of randomness in molecular orientation of the material. This will affect the tendency of permanent dipoles to align along direction of the field.

5) Hence, orientation polarization decreases with increase in temperature.

6) However in space-charge polarization, increase in temperature helps the ion movement by diffusion. As a result it will increase the polarization.

7) Thus both the orientational and space-charge polarization mechanisms are strongly temperature dependent.

4.5. Internal Field or Local Field

When a dielectric material is placed in an external electric field, it produces an induced dipole moment.

Now, two fields are acting at any point inside dielectrics are:

1) Macroscopic electrical field due to external electric field.

2) Electrical field due to electric dipole moment.

These long range coulomb fields produced due to dipoles is known as internal field or local field. This internal field is responsible for polarization of each atom or molecule in the solid.

1) *Derivation*

Lorentz Method to Find Internal Field

The dielectric material is uniformly polarized by placing it in between two plates of parallel plate capacitor as shown in figure 4.7. Assume an imaginary spherical cavity around an atom for which the internal field must be calculated at its centre.

The internal field (E_{int}) at the atom site is considered to be made up of the following four components. E_1, E_2, E_3 and E_4.

$$E\ int = E1 + E2 + E3 + E4$$

Where, E_1 – Electrical field due to charges on the plates of the capacitor.

E_2 – Electric field due to polarized charges (induced charges) on the plane surface of the dielectric.

E_3 – Electric field due to polarized charges induced on the surface of the imaginary spherical cavity.

E_4 – Electric field due to permanent dipoles of atoms inside the spherical cavity considered.

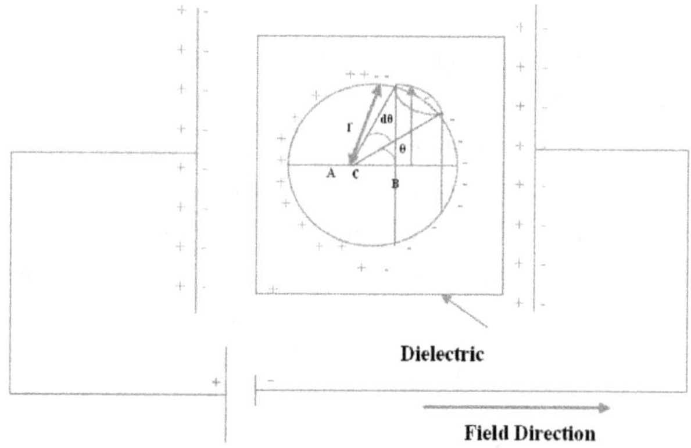

Figure 4.7: Representation to Determine Internal Field

Macroscopically we can take $E = E_1 + E_2$. i.e., the electrical field externally applied (E_1) and the electrical field induced on the plane surface of the dielectric (E_2) can be considered as a single electrical field. If we consider a dielectric that is highly symmetric, the electrical field due to dipoles present inside the imaginary cavity will cancel out of each other. Therefore the electrical field due to permanent dipoles $E_4 = 0$

Now, the equation (1) is rewritten as

$$E_{int} = E + E_3$$

Calculation of E₃

Let us consider small area ds on the surface of the spherical cavity. It is confined within an angle dθ at an angle θ in the direction of electric field E.

Polarization P is parallel to E. P_N is the component of polarization perpendicular to the area ds as shown in the figure 4.8.

$$PN = P\cos\theta$$

q' is the charge on the area ds. Polarization is also defined as the surface charges per unit area. q' / ds

$$P_N = P\cos\theta = (q' / ds)$$

Electric field intensity at C due to charge q' (Coulomb force) is given by

$$E = \frac{q'}{4\pi\varepsilon_0 r^2}$$

Substituting for q', we have

$$E = \frac{P\cos\theta\, ds}{4\pi\varepsilon_0 r^2}$$

This electrical field intensity is along the radius r and it can be resolved in to two components (E_x and E_y) as shown in figure.

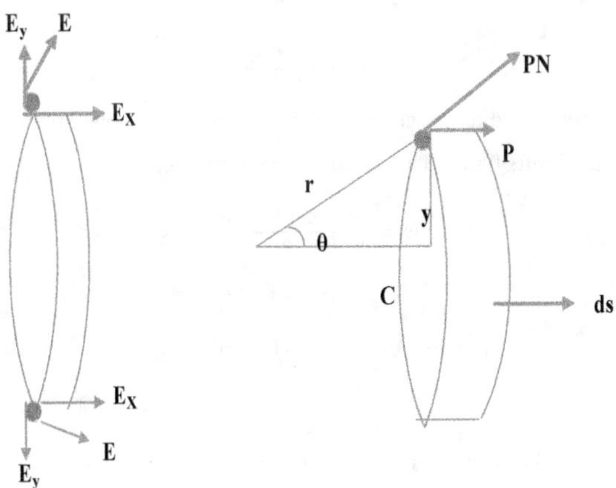

Figure 4.8: Representation of Small Area Ds

The component of intensity parallel to the electrical field direction,

$$E_x = E \cos\theta$$

Substituting for E, we have

$$E_x = \frac{P\cos\theta\cos\theta\ ds}{4\pi\varepsilon_0\ r^2}$$

$$E_x = \frac{P\cos^2\theta\cos\theta\ ds}{4\pi\varepsilon_0\ r^2}$$

The component of intensity perpendicular to the field direction,

$$E_x = E \sin\theta$$

Since the perpendicular components are in opposite directions, they cancel out each other. Hence, the parallel components alone are taken onto consideration.

Now, consider a ring area dA which is obtained by revolving ds about AB as shown in figure 4.8.

$$\text{Ring area dA} = \text{Circumference x thickness}$$

$$= 2\pi y \times rd\theta$$

$$= 2\pi\ r\ \sin\theta \times rd\theta \quad dA = 2\pi\ r2\ \sin\theta\ d\theta$$

Electric field intensity,

$$\text{Due to elemental ring dA} = \frac{P\cos^2\theta\ dA}{4\pi\varepsilon_0\ r^2}$$

By Substituting,

$$E = \frac{P\cos^2\theta\ dA}{4\pi\varepsilon_0\ r^2}\ X\ 2\pi r^2\sin\theta d\theta$$

$$E = \frac{P\cos^2\theta\sin\theta d\theta}{2\varepsilon_0}$$

Electric field intensity due to charges present in the whole sphere is obtained by integration within the limits 0 to π. This electrical field is taken as E_3.

$$E_3 = \int_0^\pi \frac{P\cos^2\theta\sin\theta d\theta}{2\varepsilon_0}$$

$$E_3 = \frac{P}{2\varepsilon_0}\int_0^\pi\cos^2\theta\sin\theta d\theta$$

$$E_3 = \frac{P}{2\varepsilon_0}\times\frac{2}{3}\quad[2/3 =\int_0^\pi\cos^2\theta\sin\theta d\theta]$$

$$E_3 = \frac{P}{3\varepsilon_0}$$

Thus the Internal field or Lorentz field, $E_{int} = E + E_3$

$$E_{int} = E + \frac{P}{3\varepsilon_0}$$

Deduction of Clausius–Mosoti Equation

Consider the dielectric material having cubic structure. Since, there are no of ions and permanent dipoles.

$$\text{Thus, } \alpha_i = \alpha_o = 0$$

For electronic polarization,

$$P = N\mu e$$

$$P = N \alpha e \, Ei$$

$$P = N \alpha e \left[E + \frac{P}{3\varepsilon_0}\right]$$

$$P = N \alpha e \, E + \frac{N \alpha_e P}{3\varepsilon_0}$$

$$P - \frac{N \alpha_e P}{3\varepsilon_0} = N \alpha_e E$$

$$P \left[1 - \frac{N \alpha_e}{3\varepsilon_0}\right] = N \alpha_e E$$

$$P = \frac{N \alpha_e E}{\left[1 - \frac{N \alpha_e}{3\varepsilon_0}\right]}$$

Since, $D = P + \varepsilon_0 E$

$$P = D - \varepsilon_0 E$$

$$P = \varepsilon E - \varepsilon_0 E$$

$$P = (\varepsilon - \varepsilon_0) E$$

$$P = (\varepsilon_r \varepsilon_0 - \varepsilon_0) E$$

$$P = E \varepsilon_0 (\varepsilon_r - 1)$$

By Comparing P,

$$\frac{N \alpha_e E}{\left[1 - \frac{N \alpha_e}{3\varepsilon_0}\right]} = E \varepsilon_0 (\varepsilon_r - 1)$$

$$\frac{N \alpha_e E}{E \varepsilon_0 (\varepsilon_r - 1)} = 1 - \frac{N \alpha_e}{3\varepsilon_0}$$

$$\frac{N \alpha_e}{3\varepsilon_0} \left[1 + \frac{3}{(\varepsilon_r - 1)}\right] = 1$$

$$\frac{N \alpha_e}{3\varepsilon_0} = \frac{1}{\left[1 + \frac{3}{(\varepsilon_r - 1)}\right]}$$

$$\frac{N \alpha_e}{3\varepsilon_0} = \frac{1}{\left[\frac{\varepsilon_r - 1 + 3}{(\varepsilon_r - 1)}\right]}$$

$$\frac{N \alpha_e}{3\varepsilon_0} = \frac{\varepsilon_r - 1}{\varepsilon_r + 2}$$

The above relation is known as Clausius–Mosoti Equation.

4.6. Dielectric Breakdown

When a dielectric is placed in an electric field and if the electric field is increased, when the field exceeds the critical field, the dielectric loses its insulating property and becomes conducting.

i.e., large amount of current flows through it. This phenomenon is called dielectric breakdown.

The electric field strength at which the dielectric breakdown occurs is known as dielectric strength.

$$\text{The dielectric strength} = \frac{\text{Dielectric Voltage}}{\text{Thickness of Dielectric}}$$

4.6.1. Types of Dielectric Breakdown

1) Intrinsic or avalanche breakdown
2) Thermal breakdown
3) Avalanche Breakdown
4) Chemical and electrochemical breakdown
5) Discharge breakdown
6) Defect breakdown

1) Intrinsic Breakdown

When dielectric is subjected to electric field then the electrons in the valance band acquire sufficient energy and go to conduction band by crossing the energy gap and hence become conduction electrons.

Therefore large current flows and it is called intrinsic breakdown or zener breakdown.

2) Avalanche Breakdown

These conduction electrons on further application of field then collide with the valance electrons in the co-valent band and remove more electrons hence transferring them as conduction electrons.

These secondary conduction electrons again dislodge some other bound electrons in the valance band and this process continues as a chain reaction.

Therefore very large current flows through the dielectrics and hence called as avalanche breakdown.

Characteristics

1) It can occur at lower temperatures.

2) It requires relatively large electric fields.

3) This kind of breakdown occurs in thin samples.

4) It occurs within short span of time

3) Thermal Breakdown

In general, when a dielectric is subjected to an electric field, heat is generated. This generated heat is dissipated by the dielectric. In some cases the heat generated will be very high compared to the heat dissipated. Under this condition the temperature inside the dielectric increases and heat may produce breakdown. This type of breakdown known as thermal breakdown.

Characteristics

1) It occurs at higher temperatures.

2) It requires moderate electric fields.

3) It depends on the size and shape of the dielectric material.

4) It occurs in the order of milliseconds.

4) Chemical and Electrochemical Breakdown

This type of breakdown is almost similar to the thermal breakdown. If the temperature is increased mobility of ions will increase and hence the electrochemical reaction may be induced to take place. Therefore when mobility of ions increased, insulation decreases and hence dielectrics becomes conducting. This type of breakdown is called as chemical and electrochemical breakdown.

Characteristics

1) It occurs only at low temperatures.

2) It depends on concentration of ions, magnitude of leakage current.

3) It occurs even in the absence of electric field.

5) Discharge Breakdown

Discharge breakdown occurs when a dielectric contains occluded air bubbles as shown in figure 4.9. When this type of dielectric subjected to electrical field, the gases present inside the material will easily ionize and thus produce large ionization current. This is known as discharge breakdown.

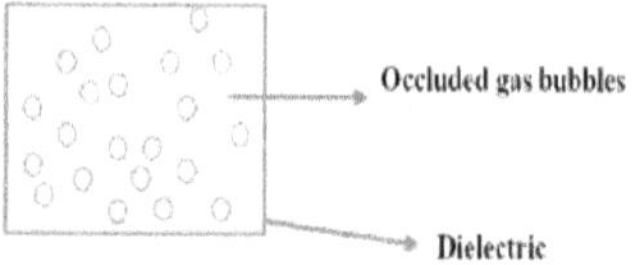

Figure 4.9: Representation of Discharge Break Down

Characteristics

1) It occurs at low voltages.

2) It occurs due to the presence of occluded air bubbles.

3) It depends upon the frequency of the applied voltage.

6) Defect Breakdown

Some dielectric have defects such as cracks, pores, blow holes etc. these vacant position may have moisture which leads to breakdown called as defect breakdown.

Remedies for Breakdown Mechanisms

1) To avoid breakdown, the dielectric material should have the following properties.

2) It should have high resistivity.

3) It must possess high dielectric strength.

4) It should have sufficient mechanical strength.   Dielectric loss should be low.

5) Thermal expansion should small.

6) It should resistive to oils, liquids and gases.

7) It must have less density.

8) There should not be any defects. It must be in pure form.

Tissue papers and polypropylene films with dielectric are used in power capacitors

General Applications

The following are the some of the applications of the dielectric materials:

1) Quartz crystal is used for the preparation of ultrasonic transducers, crystal oscillators, delay lines, filters etc.

2) Barium Titanate is used for the preparation of accelerometers.

3) Lead Zirconate Titanate ($PbZr_x Ti_{1-x} O_3$) is used for the preparation of earphones,

microphones, spark generators (gas lighter, car ignition), displacement transducer, accelerometers etc.

4) Mica is used in electrical machines, switch gears, armature winding, hot plates etc.

5) Dielectric materials are used as an insulating material in power cables, signal cables, electric motors, circuit breakers etc.

6) In radiation detectors, thermionic valves and electric devices, the dielectric materials are used.

7) The pyroelectric materials are used as radiation detector.

8) The dielectric materials are used in strain gauges, capacitors and resistors.

9) The electro-optic devices are prepared using dielectric material.

4.7. Applications of Dielectric Materials

Almost all electrical devices depend on insulating material in some way or other. Insulating materials are used in power and distribution transformers, rotating machines, capacitors, cables, and electronic applications.

1) Dielectrics in Capacitors

For dielectrics used in capacitors, it should possess the following properties.

1) It must have high dielectric constant.

2) It should possess high dielectric strength.

3) It should have high specific resistance.

4) It should also have low dielectric loss.

Uses

1. Thin sheets of papers filled with synthetic oils are used as dielectrics in the capacitors.

2. Mica used as dielectrics in discrete capacitor.

3. An electrolytic solution of sodium phosphate is used in wet type electrolytic capacitors.

4. Ceramic materials such as barium titanate and calcium titanate are used in disc capacitors and high frequency capacitors respectively.

2) Insulating Materials in Transformers

For dielectrics to act as insulating materials, it should possess the following properties.

1) It should have low dielectric resistant.

2) It should possess low dielectric loss.

3) It must have high resistance.

4) It must possess high dielectric strength.

5) It must have high moisture resistance.

6) It should have adequate chemical stability.

Uses

1) Ceramics and polymers are used as insulators.

2) Paper, rubber, plastics, waxes etc are used to form thin films, sheets, tapes, rods, etc. PVC, is used to manufacture pipes, batteries, cables etc.

3) Glass, mica, asbestos, alumina are used in ceramics.

4) Liquid dielectrics such as petroleum oils, silicone oils are widely used in transformers, circuit breakers, etc.

Synthetic oils such as askarels, sovol, etc are used as coolent and insulant in high voltage transformers.

4.8. Ferro-Electricity and Applications

1) Ferro–Electricity

When a dielectric material exhibits electric polarization even in the absence of external field, it is known as ferro-electricty and these materials termed as ferro-electric.

2) Ferro-Electrics

Ferro-electrics are anisotropic crystals which exhibit spontaneous polarization, i.e. they exhibit polarization even in the absence of external electric field.

Examples

1) Rochelle salt.

2) Potassium niobate.

3) Lithium tentalate.

4) Ammonium dihydrogen phosphate.

5) Potassium dihydrogen phosphate.

Properties

1) The dielectric constant of these Ferro-electric materials is above 2000 and it will not vary with respect temperature.

2) The dielectric constant (ε_r) reaches maximum value only at a particular temperature

called Curie temperature.

3) The polarization does not varies linearly with respect to electric field and hence these materials are also called as non-linear dielectrics.

4) Ferro-electric exhibits electric polarization easily, even in the absence of external electric field. They exhibit domain structure similar to that of a Ferro-electric material.

5) Ferro-electric materials also exhibit hysteresis, similar to that of ferromagnetic materials.

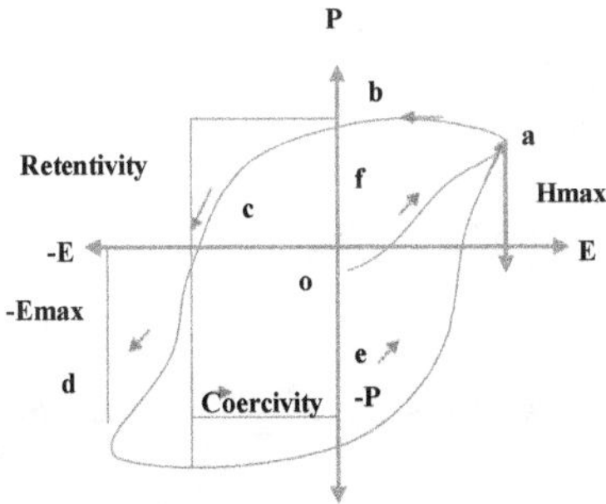

Figure 4.9.1: Hysteresis Nature of Ferro – Electric Materials

3) *Applications*

1) Ferro-electric materials are used to produce ultasonics

2) They are used in production of piezo-electric materials and in turn to make micro phones. Ferro-electrics are also used in SONAR, strain gauges, etc.

3) Ferro-electric semiconductors are used to make positors, which is turn are used to measure and control the temperature.

4) They are also used as frequency stabilizers and crystal controlled oscillators.

5) Electrets are a type of ferro-electric materials, used in the production of capacitor microphones, gas filters, etc.

6) Electrets are also used to bond the fractured bones in the human body.

7) Pyro-electric materials are also used to produce high sensitive infrared detectors.

UNIT-V

ADVANCED ENGINEERING MATERIALS

Abstract

Learning objectives. After studying this unit, one can acquire,

1) Acquire knowledge on smart materials.
2) Understand metallic glasses and its applications.
3) Understand shape memory alloys and its applications.
4) Acquire knowledge on nano particles and nano materials, their properties and applications.
5) Acquire knowledge on carbon nanotubes.

5.1. Introduction

Development of new materials has played an important role in the advancement of civilizations.

Materials like metals and their alloys, ceramics, glasses, polymers, semiconductors, biomaterials and composite materials have created a revolution in the fields of transportation, construction, energy, electronics, space technology and medicine.

In this chapter, we are going to discuss the new engineering materials like metallic glasses, shape memory alloys, etc., along with their properties and its wide range of applications.

5.2. Metallic Glasses

Metallic glass = Amorphous metal

The Metallic glasses are materials which have the properties of both metals and glasses.

In general, metallic glasses are strong, ductile, malleable, opaque and brittle.

They also have good magnetic properties and high corrosion resistance.

5.2.1. Method of Preparation

Principle

The principle used in making metallic glasses is extreme rapid cooling of the molten alloy. The technique is called as rapid quenching. The cooled molten alloys are fed into highly conducting massive rollers at high speeds to give ribbons of metallic glasses.

Melt Spinning System

A melt spinner consists of a copper roller over which a refractory tube with fine nozzle is placed. The refractory tube is provided with induction heater as shown in figure 5.1.

The metal alloy is melted by induction heating under inert gas atmosphere (helium or argon). The properly super heated molten alloy is ejected through the fine nozzle at the bottom of the refractory tube.

The molten alloy falls on the copper roller which is rotated at high speed. Thus, the alloy is suddenly cooled to form metallic glass. In this method a continuous ribbon of metallic glass can be obtained.

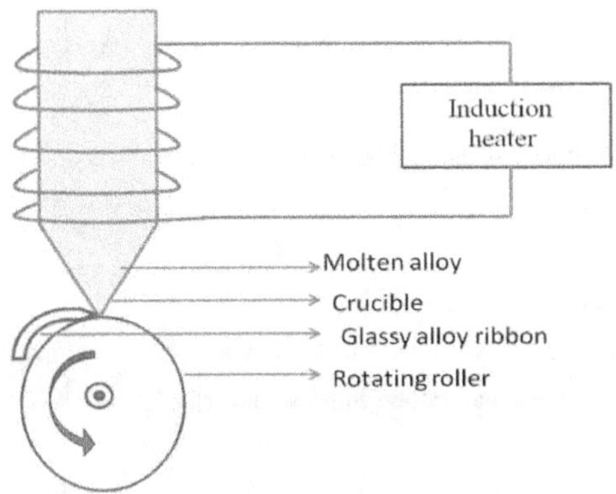

Figure 5.1: Melt Spinning Technique

5.2.2. Types of Metallic Glasses

Metallic glasses are classified into two types:

1) Metal–Metal Metallic Glasses

They are combination of metals.

Examples

Metals	Metals
Nickel (Ni)	-Niobium (Nb)
Magnesium (Mg)	-Zinc (Zn)
Copper (Cu)	-Zirconium (Zr)

2) Metal – Metalloid Metallic Glasses

These are combinations of metals and metalloids.

Examples

Metals   Metalloids

Fe, Co, Ni   -   B, Si, C, P

5.2.3. Properties of Metallic Glasses

1) Structural Properties

1) They do not have any crystal defects such as grain boundaries, dislocation etc.
2) Metallic glasses have tetrahedral close packing (TCP).

2) Mechanical Properties

1) Metallic glasses have extremely high strength, due to the absence of point defects and dislocation.
2) They have high elasticity.
3) They are highly ductile.
4) Metallic glasses are not work-harden but they are work – soften. (work harnening is a process of hardening a material by compressing it).

3) Electrical Properties

1) Electrical resistivity of metallic glasses is high and it does not vary much with temperature.
2) Due to high resistivity, the eddy current loss is very small.
3) The temperature coefficient is zero or negative.

4) Magnetic Properties

1) Metallic glasses have both soft and hard magnetic properties.
2) They are magnetically soft due to their maximum permeabilities and thus they can be magnetized and demagnetized very easily.
3) They exhibit high saturation magnetization.
4) They have less core losses.
5) Most magnetically soft metallic glasses have very narrow hysteresis loop with same crystal composition. This is shown in figure 5.2.

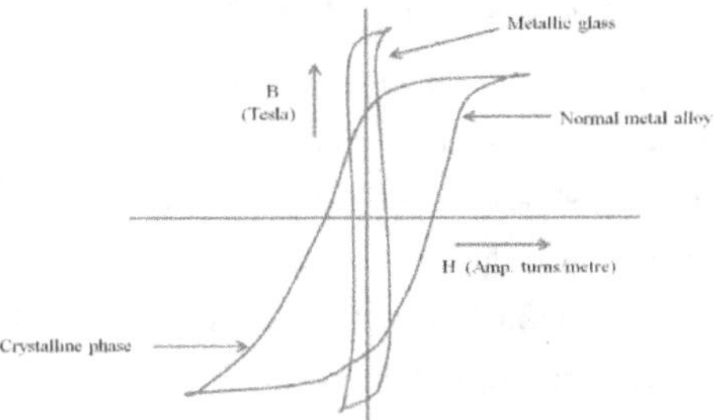

Figure 5.2: Hysteresis Loop of Iron Based Alloy in Crystalline and Metallic Glassy Phase

5) *Chemical Properties*

1) They are highly resistant to corrosion due to random ordering.

2) They are highly reactive and stable.

3) They can act as a catalyst. The amorphous state is more active than the crystalline state from the catalytic point of view.

5.2.4. *Applications of Metallic Glasses*

Metallic glasses also called as met glasses have found wide applications in different fields.

1) *Structural Application*

1) They posses high physical and tensile strength. They are superior to common steels and thus they are very useful as reinforcing elements in concrete, plastic and rubber.

2) Strong ribbons of metallic glasses are used for simple filament winding to reinforce pressure vessels and to construct large fly wheels for energy storage.

3) Due to their good strength, high ductility, rollability and good corrosion resistance, they are used to make razor blades and different kinds of springs.

2) *Electrical and Electronics*

1) Since metallic glasses have soft magnetic properties, they are used in tape recorder heads, cores of high-power transformers and magnetic shields.

2) They use of metallic glasses in motors can reduce core loss very much when compared with conventional crystalline magnets.

3) Superconducting metallic glasses are used to produce high magnetic fields and magnetic levitation effect.

4) Since metallic glasses have high electrical resistance, they are used to make accurate standard resistance, computer memories and magneto resistance sensors.

3) *Metallic Glasses as Transformer Core Material*

1) Metallic glasses have excellent magnetic properties. When they are used as transformer core, they give maximum magnetic flux linkage between primary and secondary coils and thus reduce flux leakage losses. In view of their features like small thickness, smaller area, light weight, high resistivity, soft magnetic property and negligible hysteresis and eddy current loss, metallic glasses are considered as suitable core materials in different frequency transformers.

4) *Nuclear Reactor Engineering*

1) The magnetic properties of metallic glasses are not affected by irradiation and so they are useful in preparing containers for nuclear waste disposal and magnets for fusion reactors.

2) Chromium and phosphorous based (iron chromium, phosphorous-carbon alloys) metallic glasses have high corrosion resistances and so they are used in inner surfaces of reactor vessels, etc.

5) *Bio-medical Industries*

1) Due to their high resistance to corrosion, metallic glasses are ideal materials for making surgical instruments.

2) They are used as prosthetic materials for implantation in human body.

5.3.   Shape Memory Alloys

A group of metallic alloys which shows the ability to return to their original shape or size (i.e., alloy appears to have memory) when they are subjected to heating or cooling are called shape memory alloys.

Phase of Shape Memory Alloys

Martensite and austenite are two solid phases in SMA as shown in figure 5.3.

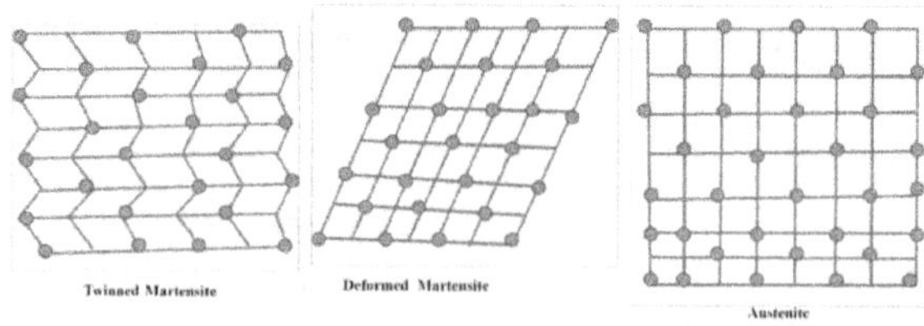

Figure 5.3: Phases of SMA

1) Martensite is relatively soft and it is easily deformable phase which exists at low temperature (monoclinic).

2) Austenite is a phase that occurs at high temperature having a crystal structure and high degree of symmetry (cubic).

5.3.1. Types of Shape Memory Alloys

There are two types of shape memory alloys:

1) One-way shape memory alloy

2) Two-way shape memory alloy

A material which exhibits shape memory effect only upon heating is known as one-way shape memory alloy. A material which shows a shape memory effect during both heating and cooling is called two-way shape memory alloy.

Examples of Shape Memory Alloys

Generally, shape memory alloys are intermetallic compounds having super lattice structures and metallic-ionic-covalent characteristics. Thus, they have the properties of both metals and ceramics.

Ni – Ti alloy (Nitinol)

Cu – Al – Ni alloy

Cu – Zn – Al alloy

Au – Cd alloy

Ni – Mn – Ga and Fe based alloys

5.3.2. Characteristics of SMA

Shape Memory Effect

1) The change of shape of a material at low temperature by loading and regaining of original shape by heating it, is known as shape memory effect.

2) The shape memory effect occurs in alloys due to the change in their crystalline structure with the change in temperature and stress.

3) While loading, twinned martensite becomes deformed martensite at low temperature.

4) On heating, deformed martensite becomes austenite (shape recovery) and upon cooling it gets transformed to twinned martensite (figure 5.4).

   a) SMAs exhibit changes in electrical resistance, volume and length during the transformation with temperature.

   b) The mechanism involved in SMA is reversible (austenite to martensite and vice versa.)

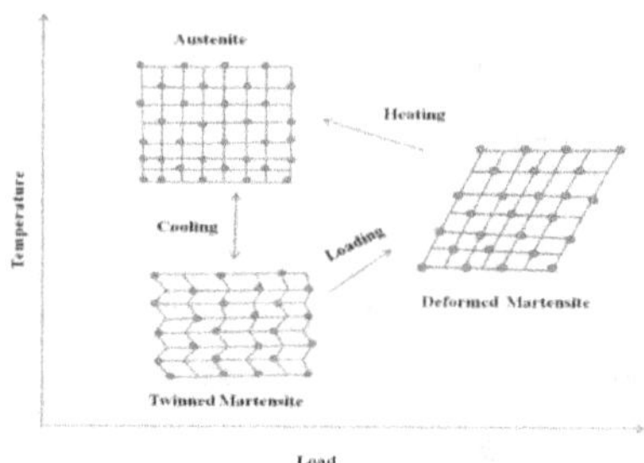

Figure.5.4: Material Crystalline Arrangement During Shape Memory Effect

1) Stress and temperature have a great influence on martensite transformation.
2) Pseudo elasticity

Pseudo–elasticity occurs in shape memory alloys when it is completely in austenite phase (temperature is greater than A_f austenite finish temperature).

Unlike the shape memory effect, Pseudo-elasticity occurs due to stress induced phase transformation without a change in temperature. The load on the shape memory alloy changes austenite phase into martensite (Figure 5.5).

As soon as the loading decreases the martensite begins to transform to austenite.

This phenomenon of deformation of a SMA on application of large stress and regaining of shape on removal of the load is known as pseudo elasticity.

This phenomenon of pseudo elasticity is also known as super elasticity.

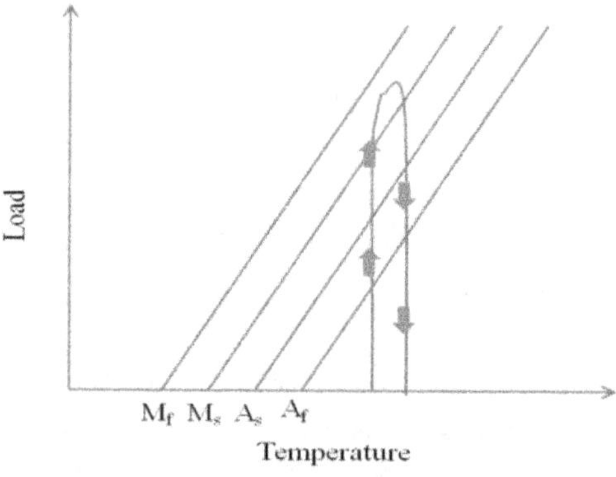

Figure 5.5: Load Diagram of Pseudo Elastic Effect

Hysteresis

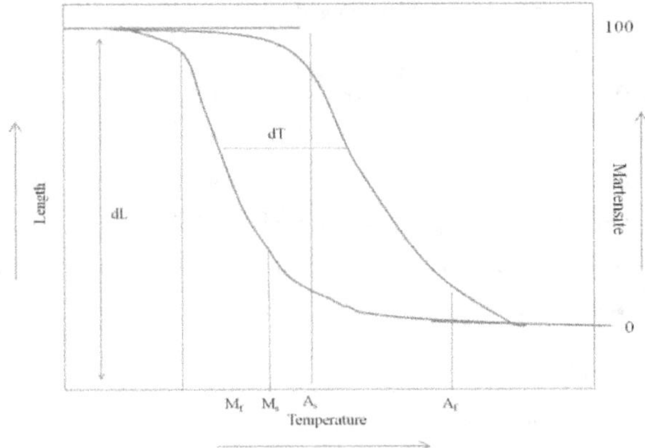

Figure 5.6. Hysteresis Curve for SMA's

The temperature range for the martensite to austenite transformation which takes place upon heating is somewhat higher than that for the reverse transformation upon cooling. The difference between the transition temperature upon heating and cooling is called hysteresis. The hysteresis curve for SMAs is shown in figure 5.6. The difference of temperature is found to be $20\text{-}30^\circ C$.

5.3.3.   *Commercial Shape Memory Alloys*

The only two alloy systems that have achieved any level of commercial exploitation are:

1) Ni-Ti alloys, and
2) Copper base alloys.

Nickel-Titanium Alloys

The basis of the Nickel-Titanium alloy is the binary, equi-atomic inter-metallic compound of Ti-Ni. The inter-metallic compound is extraordinary because it has moderate solubility range for excess Nickel or Titanium, as well as most other metallic elements. This solubility allows alloying with many of the elements to modify both the mechanical properties and the transformation properties of the system. Excess Nickel strongly depresses the transformation temperature and increases the yield strength of the austenite. The contaminants such as Oxygen and Carbon shift the transformation temperature and degrade the mechanical properties. Therefore, it is also desirable to minimize the amount of such elements.

Properties

1) The Ni-Ti alloys have greater shape memory strain up to 8.5% tend to be much more thermally stable.

2) They have excellent corrosion resistance and susceptibility, and have much higher ductility.

3) Machining by turning or milling is very difficult except with special tools.

4) Welding, brazing or soldering the alloys is generally difficult.

5) The materials do respond well to abrasive removal such as grinding, and shearing.

6) Punching can be done if thicknesses are kept small.

Advantages of Shape Memory Alloys

1) They are simple, compact and high safe.

2) They have good bio–compatibility.

3) They have diverse applications and offer clean, silent and spark-free working condition

4) They have good mechanical properties and are strong corrosion-resistant.

Disadvantages of Shape Memory Alloys

1) They have poor fatigue properties.

2) They are expensive.

3) They low energy efficiency.

5.3.4.   Applications of Shape Memory Alloys

1) Microvalve (Actuators)

One of the most common applications of SMAs is micro valves made of Ni–Ti alloy actuator. Actuator is a microsensor that can trigger the operation of a device. The electrical signal initiates an action. When an electrical current of 50 to 150 mA flows in Ni-Ti actuator, it contracts and lifts the poppet from the orifice and opens the valve.

2) Toys and Novelties

Shape memory alloys are used to make toys and ornamental goods. A butterfly using SMA. Moves its wings in response to pulses of electricity.

3) Medical Field

1) Blood clot filters are SMAs, properly shaped and inserted in veins to stop the passing blood clots. When the SMA is in contact with the clot at a lower temperature, it expands and stops the clot and blood passes through the veins.

2) They are used in artificial hearts.

3) NiTi wire holds the teeth tight with a constant stress irrespective of the strain produced by the teeth movement. It resists permanent deformation even if it is bent. NiTi is non- toxic and non-corrosive with body fluid.

4) SMAs (NiTi) are used to make eye glass frames and medical tools. Sun-glasses made from superelastic Ni-Ti frames provide good comfort and durability.

4) Antenna Wires

The flexibility of superelastic Ni–Ti wire makes it ideal for use as retractable antennas.

5) Thermostats

SMAs are used as thermostat to open and close the valves at required temperature.

6) Cryofit Hydraulic Couplings

SMAs materials are used as couplings for metal pipes.

7) Springs, Shock Absorbers, and Valves

Due to the excellent elastic property of the SMAs, springs can be made which have varied industrial applications. Some of them are listed here.

Engine micro valves, Medical stents (Stents are internal in plant supports provided for body organs), Fire safety valves and Aerospace latching mechanisms.

8) Stepping Motors

Digital SMA stepping motors are used for robotic control.

1) Titanium-aluminium shape memory alloys offer excellent strength with less weight and dominate in the aircraft industry.

2) They are high temperature SMAs, for possible use in aircraft engines and other high temperature environments.

5.4. Nanotechnology

5.4.1. Nano Materials

Nanoparticles are the particles that have three dimensional nanoscale, the particle is between 1 and 100 nm in each spatial dimension. A nanometer is a unit of measure equal to one-billionth of a meter, or three to five atoms across.Nanotechnology is the design, fabrication and use of nanostructured systems, and the growing, assembling of such systems either mechanically, chemically or biologically to form nanoscale architectures, systems and devices.

1) Comparison of Different Objects

1)	Diameter of sun	-	1,393,000km
2)	Diameter of earth	-	1,28,000km
3)	Height of Himalaya mountain	-	8,848km
4)	Height of man	-	1.65km
5)	Virus	-	20-250nm
6)	Cadmium sulphide nanoparticle	-	1-10nm

2) Classification of Nanomaterials

1) Clusters - A collection of atoms or reactive molecules up to about 50 units.

2) Colloid - A stable liquid phase containing particles in 1 to 1000 nm range. A colloidal particle is one such 1 to 1000 nm sized particle.

3) Nanoparticle - A solid particle in the 1 to 100 nm range that could be non-crystalline, an aggregate of crystallites, or a single crystallite.

4) Nanocrystal - A solid particle that is a single crystal in the nanometer size.

5) Nanostructured or Nanoscale Material - any solid materials have a nanometer dimension.

6) Three dimensions   - Particles

    Two dimensions   - Thin films

    One dimension   - Thin wire

7) Quantum Dots

A particle that exhibits a size quantization effect in at least one dimension.

5.4.2. Synthesis Processes

Nano materials are newly developed materials with grain size at the nanometre range (10^{-9} m) i.e., in the order of 1 – 100 nm. The particle size in a nano material is in the order of nm.

1) Top-down Process

In this processes, bulk materials are broken into nano sized particles as represented in figure below.

In top-down processes, the building of nanostructures starting with small components like atoms and molecules that are removed from a bulk material so as to obtain desired microstructure.

2) Bottom-up Processes

In this processes, nano phase materials are produced by building of atom by atom as represented in figure below.

This processes building larger objects from smaller buildings blocks. Nanotechnology seeks to use atoms and molecules as those building blocks. This is the opposite of the top-down approach. Instead of taking material away to make structures, the bottom-up approach selectively adds atoms to create structures.

1. Pulsed Laser Deposition

Principle

The laser pulse of high intensity and energy is used to evaporate carbon from graphite. These evaporated carbon atoms are condensed to from nanotubes.

Description

The experimental arrangement of pulsed laser4 deposition is shown in figure 5.7. A quartz tube which contains a graphite target is kept inside a high temperature muffle furnace.

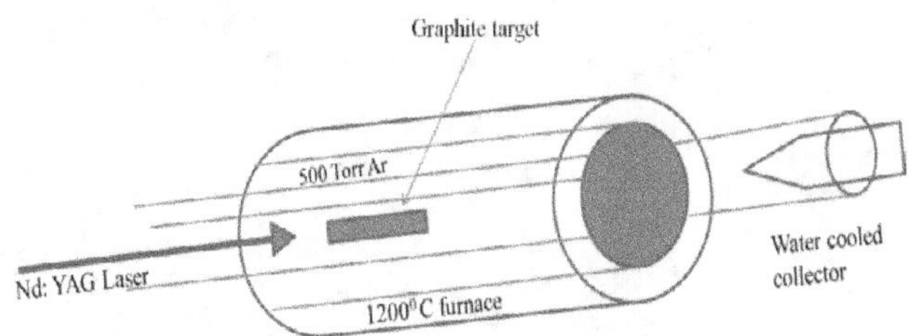

Figure 5.7: Representation of Pulsed Laser Deposition Technique

This quartz tube is filled with argon gas and it is heated to 1473 K. A water cooled copper collector is fitted at the other end of the tube. The target material graphite contains small amount of nickel and cobalt as a catalyst to nucleate the formation of nanotubes.

Working

When an intense pulse of laser beam is incident on the target, it evaporates the carbon from the graphite. The evaporated carbon atoms are swept from the higher temperature argon gas to the colder copper collector.

When the carbon atoms reach the colder copper collector, they condense into nanotubes.

2. *Chemical Vapour Deposition*

The deposition of nano films from gaseous phase by chemical reaction on high temperature is known as chemical vapour deposition. This method is used to prepare nano-powder.

Principle

In this technique, initially the material is heated to gaseous state and then it is deposited on a solid surface under vacuum condition to form nano powder by chemical reaction with the substrate.

Description and Working

The CVD reactor built to perform CVD processes is shown in figure 5.8.

Chemical vapour deposition (CVD) involves the flow of a gas with diffused reactants (substances to be deposited in the vapour) over a hot substrate surface. The gas that carries the reactants is called the carrier gas.

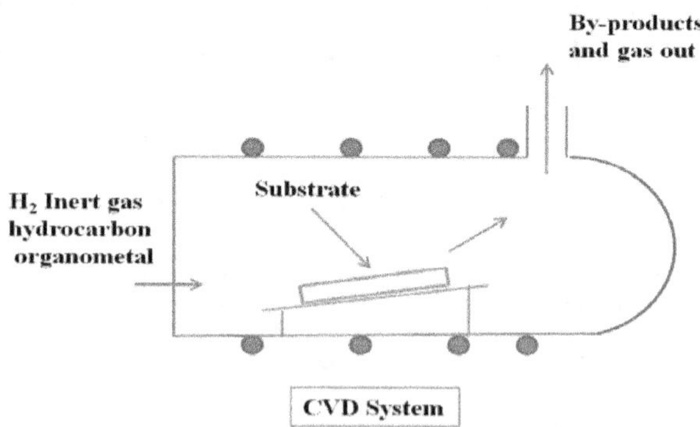

Figure 5.8: Representation of Chemical Vapor Deposition Technique

While the gas flows over the hot solid surface, the heat energy increases chemical reactions of the reactants that form film during and after the reactions.

The byproduct of the chemical reactions are then removed. The thin film of desired composition can thus be formed over the surface of the substrate.

5.4.3. *Properties of Nanophase Materials*

The mechanical, electrical, chemical, magnetic and structural properties of nanophase materials change with the reduction in the particle size of the material.

Physical Properties

Starting from the bulk, the first effect of reducing the particle size is to create more surface sites. This in turn changes surface pressure and interparticle spacing.

1) Interparticle spacing decreases with decrease in grain size for metal clusters. For example in copper, it decrease from 2.52 (cluster size – 50A°) to 2.23A° (Cu dimer) figure 5.9. The change in inter particle spacing and large surface to the volume ratio in particles have a combined effect on material properties. Therefore, the nanophase materials have very high strength and super hardness. Because of the cluster of grains, the nano phase materials are mostly free from dislocations and stronger than conventional metals.

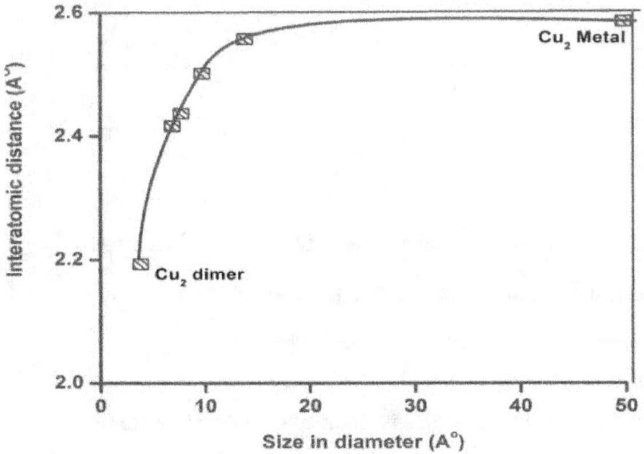

Figure 5.9: Interatomic Distance in Cun as a Function of Grain Size

2) Melting point reduces with decrease in cluster size.

   The melting point of gold in nano phase (Aun) varies as a function of particle size (figure 5.9.1.).

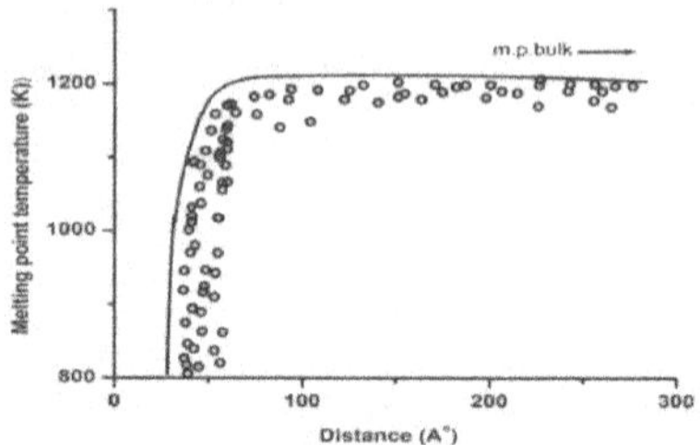

Figure 5.9.1: Melting Point of Small Aun Particles as a Function of Size

The melting point decreases from 1200 K to 800 K when the particle size decreases from 300 A° to 20 A°.

3) Ionisation potential changes with cluster size of the nanograins.

The electronic bands in metals become narrower when the size is reduced from bulk which changes the value of ionization potential.

Figure 5.9.2. shows the ionization potential and reactivity of Fen clusters as a function of size. Ionisation potentials are higher at small sizes than that for the bulk and show marked fluctuations as a function of size.

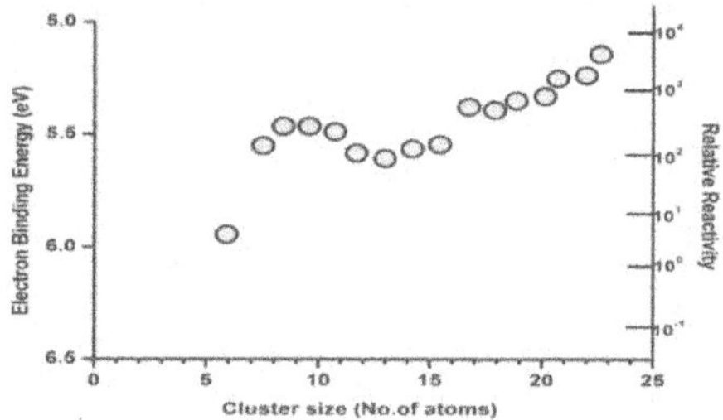

Figure 5.9.2: Ionisation Potential and Reactivity of Fen Clusters as a Function of Size

4) The large surface to volume ratio, the variations in geometry and the electronic structure have a strong effect on catalytic properties.

As an example, the reactivity of small clusters is found to vary by higher orders of magnitude when the cluster size is changed by only a few atoms.

Magnetic Properties

Nanoparticles of non-magnetic solids also exhibit totally new type of magnetic properties.

1) Bulk magnetic moment increases with decrease in co-ordination number The change in magnetic moment on the nearest coordination number is shown in figure 5.9.3. As the

coordination number decreases, the magnetic moment increases with the atomic value which means that small particles are more magnetic than the bulk material. The magnetic moment of iron (Fe) of nanoparticles is 30% more than that of bulk. At smaller sizes, the clusters become spontaneously magnetic.

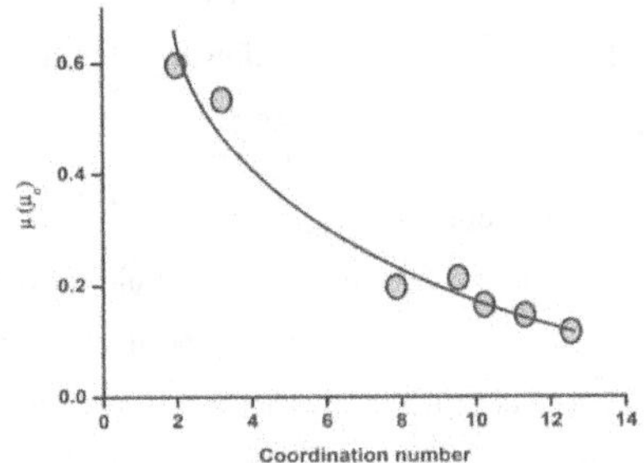

Figure 5.9.3: Change in Magnetic Moment on the Nearest Coordination Number

2) The nano-materials shows variation in their magnetic property when they change from bulk state to cluster (nano-particle) state.

3) Non-magnetic materials become magnetic when the cluster size reduces to 80 atoms.

Mechanical Properties

1) In nanophase materials, the elastic strength is low however; its plastic behavior is high.

2) In some nanophase materials, it is noted that there is decrease in hardness when the grain size is less than 10 nm.
   However for many nanocrystalline, pure metals (10 nm), the hardness is about 2 to 7 times greater than that of large-grained (>1 μ m) metals.

3) Higher hardness and mechanical strength (2-7 times) when grain size reduces from 1 μ m to 10 nm.

4) It has very high ductility and super plastic behavior at low temperatures.

5.4.4. Applications of Nanophase Materials

1) Materials Technology

1) We can synthesis harder metals having hardness 5 times higher than normal metals using nanoparticles.

2) Stronger, lighter, wear resistant, tougher and flame retardant polymers are synthesized with nanoparticles as fillers. They are used in replacement of body parts and metals (bio-materials).

3) We can produce unusual colour paints using nanoparticles since nanoparticles exhibit entirely different optical properties.

4) Nanophase materials are used in annoelectronic devices such as nanotransistore, ceramic capacitors for energy storage, noise filters and stabilizers. The special features of these devices include smaller sizes and reduced power losses.

5) ZnO thermistors are used in thermal–protection and current-controlling devices.

2) Information Technology

1) Nanoparticles are used for data storage.

2) Quantum electronic devices have started replacing bulk conventional devices.

3) Nano materials are used to produce very tiny permanent magnets of high energy products. Hence, they are used in high-density magnetic recording.

4) Magnetic devices made of Cu-Fe alloy are used in RAM, READ/WRITE heads and sensors.

5) Quantum dots, quantum wells and quantum wires are mainly produced from semiconductor nanomaterials. Hence, they are used in computer storage (memory) devices.

3) Biomedicals

1) Biosensitive nanoparticles are used for tagging of DNA and DNA chips.

2) Controlled drug delivery is possible using nanotechnology. Diffusion of medicine through nanoporous polymer reservoir as per the requirement is very useful in controlling the disease.

3) Nanostructured ceramics readily interact with bone cells and bence finds applications as an implant material.

4) Energy Storage

1) Since the hydrogen absorbing capability increases with decrese of size of nanoparticles, nanoparticles of *Ni, Pd* and *Pt* are useful in hydrogen storage devices.

2) Metal nanoparticles are very useful in fabrication of ionic batteries.

5) Optical Devices

1) Nanomaterials are used in making efficient semiconductor laser and CD's.

2) Nano particulate zinc oxide is used to manufacture effective Sunscreens.

3) Nanoparticles are used in the coatings for eye glasses to protect from scratch or breakage.

6) Transmission Lines

Nano phase materials are used in the fabrication of signal processing elements such as filters, delay lines, switches etc.

1) Nano micro-Electro Mechanical Systems (Nano MEMS) have direct implications on integrated circuits, optical switches, pressure sensors and mass sensors.

2) Molecular Nano-Technology(MNT) is aimed to develop robotic machines, called assemblers on a molecular scale, molecular-size power sources and batteries.

3) Underwater nano sensor networks are used to detect the movement of ships in an efficient manner with faster response. They can also detect chemical, biological or radiological materials in cargo containers.

5.5. Non-Linear Materials and Bio-Materials

5.5.1. Birefringence and Kerr Effect

The appearance of double refraction under the influence of an external agent is known as artificial double refraction or induced birefringence.

Optical Kerr Effect

1) Anisotropy induced in an isotropic medium under the influence of an electric field is known as Kerr effect.

2) A sealed glass cell known as Kerr cell filled with a liquid comprising of asymmetric molecules is used to study the Kerr effect.

3) Two plane electrodes are placed in parallel to each other. When a voltage is applied to there electrodes, a uniform electric field is produced in the cell.

4) The Kerr cell is placed between a crossed polarizer system (Figure 5.9.4), When the electric field is applied, the molecules of the liquid tend to align along the field direction.

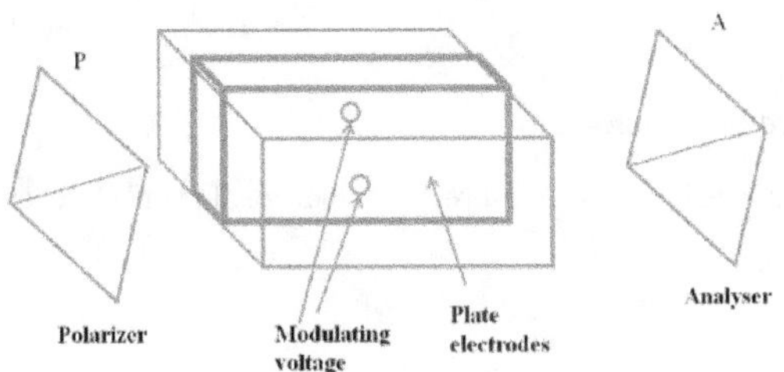

Figure 5.9.4: Kerr effect – Birefringence is Induced in a Liquid Subjected to an Electric Field

As the molecules are asymmetric, the alignment causes anisotropy and the liquid becomes double refracting. The induced birefringence is proportional to the square of the applied electric field E and to the wavelength λ of incident light. Thus

The change in refractive influx is given by

$$\Delta \mu = K \lambda E^2$$

Where K is known as the Kerr constant

5.5.2. Non-Linear Properties and Second Harmonic Generation

1) Basic Principle of Non Linear Properties

We know that a light wave is electromagnetic in nature ie., it consists of electric and magnetic fields. When the light propagates through a material, it changes the properties of the medium, such as the refractive index. It depends on the electric and magnetic fields associated with the light.

For example, we could not observe nonlinear effects with the ordinary light beam of low intensity, since the electric and magnetic fields associated with the light beams is very weak.

With the invention of laser, it is now possible to have electric fields which are strong enough to observe interesting non linear effects.

Thus if electric and magnetic fields are strong enough, the properties of the medium will be affected which in turn will affect the propagation of the light beam.

2) Non Linear Properties

Few of the nonlinear phenomena observed are

1) Second harmonic generation
2) Optical mixing
3) Optical phase conjugation
4) Soliton

3) Second Harmonic Generation

In a linear medium, polarization P is directly proportional to the electric field E

$P \propto E$.

$P = \varepsilon_0 \chi E$

Where ε_0-Permittivity of free space

χ-Electrical susceptibility

In nonlinear medium for higher fields ie., higher intensities of light the non linear effects are observed.

$$P = \varepsilon_o \left(\chi_1 E + \chi_2 E^2 + \chi_3 E^3 + \ldots \right) \tag{1}$$

Where χ_1 is the linear susceptibility and χ_2, χ_3 . . . are higher order non linear susceptibilities. The electric field passing through the medium with amplitude E_0 is

$$E = E_o \cos\omega t \tag{2}$$

Substituting value of (2) in (1) we have

$$P = \varepsilon_o \left(\chi_1 E_o \cos\omega t + \chi_2 E_o^2 \cos^2\omega t + \chi_3 E_o^3 \cos^2\omega t + \ldots \right) \tag{3}$$

We know that,

$$\cos^2\omega t = \frac{1 + \cos 2\omega t}{2} \tag{4a}$$

$$\cos^3\omega t = \frac{\cos 3\omega t + 3\cos\omega t}{4} \tag{4b}$$

Substituting equation (4 a, b) in (3), we have,

$$P = \varepsilon_o \left(\chi_1 E_o \cos\omega t + \chi_1 E_o^2 \left(\frac{1 + \cos 2\omega t}{2}\right) + \chi_3 E_o^3 \left(\frac{\cos 3\omega t + 3\cos\omega t}{4}\right) + \ldots \right)$$

$$= \varepsilon_o \left(\chi_1 E_o \cos\omega t + \chi_2 E_o^2 \left(\frac{1}{2}\right) + \chi_2 E_o^2 \left(\frac{\cos 2\omega t}{2}\right) + \chi_3 E_o^3 \left(\frac{\cos 3\omega t}{4}\right) + \chi_3 E_o^3 \left(\frac{3}{4}\cos\omega t\right) \ldots \right)$$

$$P = \varepsilon_o \left(\left(\frac{1}{2}\right) \chi_2 E_o^2 + \chi_1 E_o \cos\omega t + \chi_3 E_o^3 \left(\frac{3}{4}\cos\omega t\right) + \chi_2 E_o^2 \left(\frac{\cos 2\omega t}{2}\right) + \chi_3 E_o^3 \left(\frac{\cos 3\omega t}{4}\right) + \ldots \right)$$

$$= \varepsilon_o \left(\frac{1}{2} \chi_2 E_o^2 + \left(\chi_1 + \frac{3}{4}\chi_3 E_o^3\right) E_o \cos\omega t + \frac{1}{2} \chi_2 E_o^2 \cos 2\omega t + \frac{1}{4} \chi_3 E_o^3 \cos\omega t \right)$$

$$P = \frac{1}{2} \varepsilon_o \chi_2 E_o^2 + \varepsilon_o \left(\chi_1 + \frac{3}{4}\chi_3 E_o^3\right) E_o \cos\omega t + \frac{1}{2} \varepsilon_o \chi_2 E_o^2 \cos 2\omega t + \frac{1}{4} \varepsilon_o \chi_3 E_o^3 \cos\omega t + \ldots \tag{5}$$

In the above equation, 1st term gives rise to *dc* field across the medium, the second term gives external polarization and is called first or fundamental harmonic polarisability.

The third term which oscillates at a frequency 2w is called second harmonic of polarization and other terms are referred as higher harmonic polarization.

Both first term (dc field) and third term (second harmonic of polarization) added together is called optical rectification.

The second harmonic generation is possible only the crystals lacking inversion symmetry. SHG crystals are quartz, potassium dihydrogen phosphate (KDP), Ammonium dihydrogen phosphate (ADP), Barium titante ($BaTiO_3$) and Lithium Iodate ($LiIO_3$).

The observation of second harmonic generation by KDP is shown in figure.

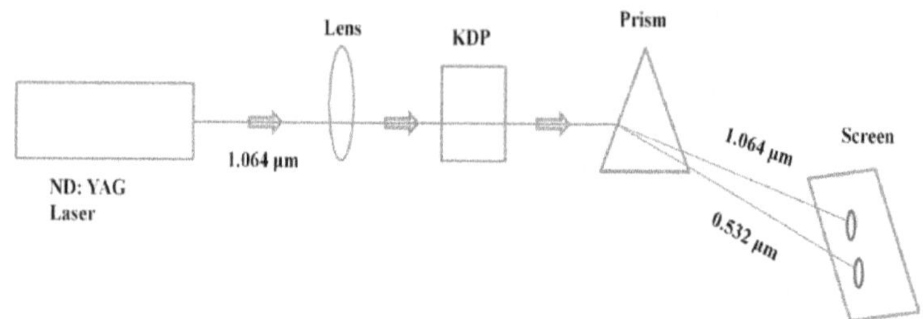

Figure 5.9.5: Arrangement for Observing Second Harmonic Generation

When the fundamental radiation (1.064 m) from Nd: YAG laser is sent through SHG crystal like KDP, conversion takes place to double the frequency. i.e., half the wavelength (0.532 m) takes place.

5.5.3. Biomaterials with their Properties and Applications

The materials which are used for structural applications in the field of medicine are known as Biomaterials. In the recent years, new biomaterials like nano biomaterials are emerging up due to the requirements in the medical field for different applications.

Classification of Biomaterials

Based on the applications in the field of medicine, biomaterials are classified as

1) Metals and alloys biomaterials
2) Ceramics biomaterials
3) Polymer biomaterials
4) Composite biomaterials

Sometimes, a single material mentioned above cannot fulfill the complete requirements imposed for specific applications. In such case, combinations of more than one material are required.

1) Metals and Alloys

Metals and alloys are used as biomaterials due to their excellent electrical and thermal conductivity and mechanical properties.

Types of Biomaterials Using Metals and Alloys

1) Cobalt based alloys
2) Titanium
3) Stainless steel

4) Protosal from cast alloy

5) Conducting metals such as Platinum

Applications

The metals and alloys biomaterials are used in implant and orthopedic applications.

1) Stainless steel is the predominant implant alloy. This is mainly due to its ease of fabrication and desirable mechanical properties and corrosion resistant.

2) Proposal from cast alloy of Co – Cr – Mo is used to make stem and used for implant hip end oprosthesis.

3) The advanced version of protosal – 10 from Co – Ni – Cr – Mo alloy is widely used in Hip joints, Ankle joints, Knee joints, leg lengthening space as.

4) ASTMF – 136 (composition of Ti – 6A1 – 4V, EL1 alloy, forged) due to its high strength/ weight ratio, high corrosion resistance and high bio compatibility, this alloy is used in dental applications for making screws, wires and artificial teeth.

5) Ni – Ti shape memory alloy is used in dental arch wires, micro surgical instruments, blood clot filters, guide wires etc.

2) Ceramics Biomaterials

Ceramics are used as biomaterials due to their high mechanical strength and biocompatibility.

Types of Bio-Ceramic Materials

1) Tricalcium phosphate.
2) Metal oxides such as Al_2O3 and SiO_2.
3) Apatite ceramics.
4) Porous ceramics.
5) Carbons and Alumina.

Applications

1) Ceramic implants such as Al_2O_3 and with some SiO_2 and alkali metals are used to make femoral head. This is made from powder metallurgical process.

2) Tricalcium phosphate is used in bone repairs.

3) Orthopedic uses of alumina consists of hip and knee joints, tibical plate, femur shaft, shoulders, radius, vectebra, leg lengthening spaces and ankle joint prosthesis. Porous alumina is also used in teeth roots.

4) Apatite ceramics are new bio active ceramics. They are regarded as synthetic bone,

readily allows bone in growth, better than currently used alumina Al_2O_3.

5) Carbon has good biocompatibility with bone and other tissues. It has high strengths and elastic molecules close to that of bone.

6) Carbon coatings find wide applications in heart valves, blood vessel grafts, percutaneous devices because of exceptional compatibility with soft tissues and blood.

7) Percutaneous carbon devices containing high density electrical connectors have been used for the chronic stimulation of the cochlea for artificial hearing and stimulation of the visual cortex to aid the blind.

3) Bio Polymers

Biopolymers are macromolecules (protein, nucleic acids and polysachacides) formed in nature during the growth cycles of all organisms.

Biopolymers find variety of applications as biomaterials. The most prominent among them are collagens, muco-polysaccharides – chitin, collagens and its derivatives.

Collagens which are major animal structural proteins are widely used in a variety of forms such as solution, gel, fibers, membranes, sponge and tubing for large number of biomedical applications including drug delivery system, vessels, valves corneal prosthesis, wound dressing, cartilage substitute and dental applications.

Biomaterials in Ophthalmology

Biomaterials find important applications in ophthalmology. They are used to improve and maintain vision. Eye implants are used to restore functionality of cornea, lens, etc, when they are damaged or diseased. The biomaterials include viscoelastic solutions intraocular lenses, contact lenses, eye shields, artificial tears, vitreous replacements, correction of corneal curvature.

Dental Materials

Polymers, composites, ceramic materials and metal alloys are four main groups of materials used for dental applications. A large number of materials are tested for porous dental implants, which include stainless steel, Co – Cr – Mo alloy, PMMA, proplast and Deacon, velour coated metallic implants, porous calcium aluminates single crystal alumina, biogas, vitreous and pyrolytic carbons.

The dental applications include impression materials, dentine base and ceorons, bridges, inlays and repair or cavities, artificial teeth, repair of alveolar bone.

Unit-I

Conducting Materials

1. **What are the merits of classical free electron theory?**

   (i) It is used to verify Ohm's law.

   (ii) It is used to explain electrical and thermal conductivities of metals.

   (iii) It is used to derive Wiedemann – Franz law.

   (iv) It is used to explain the optical properties of metal.

2. **What are the drawbacks of classical free electron theory?**

   (i) Classical theory states that all free electrons will absorb the supplied energy; on the contrary, quantum theory states that only a few electrons will absorb the supplied energy.

   (ii) Electrical conductivity of semiconductors and insulators (non–metal) could not be explained by this theory.

   (iii) The phenomena such as photo–electric effect, Compton effect and black body radiation could not be explained on the basis of this theory because these phenomena are based on quantum theory.

3. **Define mean free path**

   The average distance traveled by a free electron between any two successive collisions in the presence of an applied field is known as mean free path. It is the product of drift velocity of the electron (vd) and collision time (τ)

   $$\lambda = v_d \times \tau_c$$

4. **Define relaxation time of an electron**

   The average time taken by a free electron to reach its equilibrium position from its disturbed position due to application of an external electrical field is called relaxation time.

5. **Define drift velocity of electron. How is it different from the thermal velocity of an electron?**

   The average velocity acquired by a free electron in a particular direction after a steady state is reached on the application of an electrical field is called drift velocity. It is denoted as vd.

   The thermal velocity is random in nature and its value is very high (10^5 m/s), but the drift velocity is unidirectional and its value is very small (50 cm/s).

6. **Define mobility of electrons.**

The magnitude of the drift velocity per unit electric field is defined as the mobility of electrons (μ)

$$\text{i.e., } \mu = v_d/E$$

Where, $v_d \rightarrow$ drift velocity of electrons $E \rightarrow$ Electrical field.

7. **Define electrical conductivity.**

It is the amount of electrical charge (q) conducted per unit time (t) across unit area (A) of the solid per unit applied electrical field (E).

$$\sigma = q/tAE$$

8. **State Wiedemann – Franz law.**

It states that the ratio of thermal conductivity (K) to electrical conductivity (σ) of a metal is directly proportional to absolute temperature (T) and this ratio is constant for all metals at a given temperature.

$$K/\sigma \, \alpha \, T$$
$$\text{i.e., } K/\sigma = LT$$

Where L is a constant and it is known as Lorentz number.

9. **What is Lorentz number?**

The ratio between thermal conductivity (K) of a metal to the product of electrical conductivity (σ) of a metal and absolute temperature (T) of the metal is a constant. It is called Lorentz number and it is given by

$$L = K/\sigma \, T$$

10. **Define Fermi distribution function.**

The probability F (E) of an electron occupying a given energy level at temperature T is known as Fermi distribution function. It is given by,

$$F\,(E) = 1/1 + e^{(E-E_F)/KT}$$

K $\rightarrow$ Boltzmann's temperature

T $\rightarrow$ Absolute temperature

E $\rightarrow$ Energy of the level whose occupancy is being considered.

11. **Define Fermi level and Fermi energy with its importance.**

Fermi level is the energy level at finite temperature above 0K in which the probability of the electron occupation is ½ and it is also the level of maximum energy of the filled states at 0K

Fermi energy is the energy of the state at which the probability of the electron occupation is ½ at any temperature above 0K.It is also the maximum energy of filled states at 0K.

Importance Fermi level and Fermi energy determine the probability of an electron occupying a given energy level at a given temperature.

12. Define density of states. What is its use?

It is defined as the number of available electron states per unit volume in an energy interval E and E+dE. It is denoted by Z (E). It is used to determine Fermi energy at any temperature.

13. What are difference between Drift velocity and thermal velocity of an electron?

S.No	Drift Velocity	Thermal Velocity
1	Drift velocity is the average velocity acquired by the free electron, in the presence of electric field.	Thermal velocity is the velocity of an electron without any external field.
2.	The electrons moving with drift velocity moves in the direction opposite to that of the field	The direction of the electrons moving with thermal velocity is random.
3.	The velocity is very less, say in the order of 0.5 m/s.	The velocity is very high, say in the order of 10^5 m/s.

14. Distinguish between electrical conductivity and thermal conductivity.

S.No	Electrical Conductivity	Thermal Conductivity
1.	The co-efficient of electrical conductivity is defined as the quantity of electricity flowing per unit area per unit time maintained at unit potential gradient.	The Co-efficient of thermal conductivity is defined as the quantity of heat conducted per unit area per unit time maintained at unit temperature gradient.
2.	Electrical conductivity is purely due to number of free electrons.	Thermal conductivity is due to both free electrons and phonons.
3.	Conduction of electricity takes place from higher potential end to the lower potential end.	Conduction of heat takes place from hot end to cold end
4.	Unit: ohm^{-1} m-1	Unit: Wm^{-1} K^{-1}

Unit–II

Semiconducting Materials

1. What are elemental semiconductors? Give some important elemental semiconductors.

Elemental semiconductors are made from single element of the forth group elements of the periodic table.

It is also known as indirect band gap semiconductor.

Example: Important elemental semiconductors germanium and silicon.

2. What are the properties of semiconductors?

(i) They are formed by covalent bond.

(ii) They have empty conduction band.

(iii) They have almost filled valance band.

(iv) These materials have comparatively narrow energy gap.

3. Mention any four advantages of semiconducting materials.

(i) It can behave as insulators at 0K and as conductors at high temperature.

(ii) It possess some properties of both conductors and insulators.

(iii) On doping we can produce both N and P-type Semiconductors with charge carriers of electrons and holes respectively.

(iv) It possesses many applications in electronic field such as manufacturing of diodes, transistors, LED's, IC etc.

4. What are compound semiconductors? Give some important compound semiconductors.

Semiconductors which are formed by combining third and fifth elements or second and sixth group elements in the periodic table are called compound semiconductors.

Important compound semiconductors are

S.No	Group	Compound Semiconductor
1.	Combination of third and fifth group elements (III and V)	Gallium Phosphide (GaP) Indium Phosphide (InP) Indium Arsenide (InAs)
2.	Combination of second and sixth group elements (II and VI)	Magnesium Oxide (MgO) Magnesium Silicon (MgSi) Zinc Oxide (ZnO)

5. **What are the differences between elemental semiconductors and compound semiconductors?**

S.No	Elemental Semiconductors	Compound Semiconductors
1.	They are made of single element Eg: Ge,Si	They are made of compounds Eg: GaAs, GaP, MgO etc
2.	They are called as indirect band gap semiconductors. i.e., electron-hole recombination takes place through traps, which are present in the band gap.	They are called as direct band gap semiconductors. i.e., electron-hole recombination takes place directly with each other.
3.	Here, heat is produced during recombination.	Here, the photos are emitted during
4.	They are used for the manufacture of diodes and transistors., etc.	They are used for making LED's, laser diodes, IC's etc.

6. **What is Fermi level in a semiconductor?**

Fermi level in a semiconductor is the energy level situated in the band gap of the semiconductor. It is exactly located at the middle of the band gap in the case of intrinsic semiconductor.

7. **Define Hall-effect and Hall voltage.**

When a conductor (metal or semiconductor) carrying a current (I) is placed in a transverse magnetic field (B),a potential difference (electric field) is produced inside the conductor in a direction normal to the directions of both the current and magnetic field.

This phenomenon is known as **Hall-effect** and the generated voltage is called **Hall-voltage.**

Hall field per unit current density per unit magnetic induction is called hall coefficient.

8. **Mention the applications of Hall Effect.**

It is used to,

(i) Find type of semiconductor.

(ii) Measure carrier concentration.

(iii) Find mobility of charge carrier.

(iv) Measure the magnetic flux density using a semiconductor sample of known Hall coefficient.

9. **What is a semiconductor?**

Semiconductor is a special class of material which behaves like an insulator at o K and acts conductor at temperature other than 0K. Its resistivity lies in between a conductor and an insulator.

10. **What is an intrinsic semiconductor?**

Semiconductor in an extremely pure form (without impurities) is known as intrinsic semiconductor.

11. What is an extrinsic semiconductor?

A semiconducting material in which impurity atoms added (doped) to the material to modify its conductivity is known as extrinsic semiconductor or **impurity semiconductor.**

12. What is meant by intrinsic semiconductor and extrinsic semiconductor? What are the differences between intrinsic and extrinsic semiconductor.

S.No	Intrinsic Semiconductor	Extrinsic Semiconductor
1.	Semiconductor in a pure form is called intrinsic semiconductor.	Semiconductor which are doped with impurity is called extrinsic semiconductor
2.	Here the change carriers are produced only due to thermal agitation.	Here the change carriers are produced due to impurities and may also be produced due to thermal agitation.
3.	They have low electrical conductivity.	They have high electrical conductivity.
4.	They have low operating temperature.	They have high operating temperature.
5.	At 0K, Fermi level exactly lies between conduction band and valence band.	At 0K, Fermi level exactly lies closer to conduction band in "n" type semiconductor and lies near valence band in "p" type semiconductor.
	Examples: Si,Ge,etc.	Examples: Si and Ge doped with Al, In,P,As etc

13. What is an n-type semiconductor?

When a small amount of pentavalent impurity is added to a pure semiconductor, it becomes extrinsic or impure semiconductor and it is known as n-type semiconductor.

14. What is a p-type semiconductor?

When a small amount of trivalent impurity is added to a pure semiconductor, it becomes extrinsic or impure semiconductor and it is called p-type semiconductor.

15. What is meant by doping and doping agent?

The technique of adding impurities to a pure semiconductor is known as doping and the added impurity is called doping agent.

16. What is meant by donor energy level?

A pentavalent impurity when doped with an intrinsic semiconductor donates one electron which produces an energy level called donor energy level.

17. What is meant by acceptor energy level?

A trivalent impurity when doped with an intrinsic semiconductor accepts one electron which produces an energy level called acceptor energy level.

18. Compare n-type and p-type semiconductors.

S.No	N-type semiconductors	P-type semiconductors
1.	N-type semiconductor is obtained by doping and intrinsic semiconductor with pentavalent impurity.	P-type semiconductor is obtained by doping and intrinsic semiconductor with trivalent impurity.
2.	Here electrons are majority carriers and holes are minority carriers.	Here holes are majority carriers and electrons are minority carriers
3.	It has donor energy levels very close to CB	It has acceptor energy levels very close to VB.
4.	When the temperature is increased, these semiconductors can easily donate an electron from donor energy level to the CB.	When the temperature is increased, these semiconductors can easily accept an electron from VB to donor energy level .

19. Define impurity range, exhaustion range and intrinsic range in n-type Semiconductors.

Impurity range: This range is due to the transfer of electrons from the donor energy Level to CB. Here the electron concentration in the CB steadily increases due to ionization of donor atoms.

Exhaustion range: When all the electrons are transferred from donor energy level to conduction band, the electron concentration remains constant over certain temperature and is called exhaustion range.

Intrinsic range: In this range the n-type semiconductor practically behaves like the Intrinsic semiconductor. Therefore if the temperature is increased the electrons concentration in the conduction band increases rapidly due to the shifting of electrons from valence band to conduction band.

Unit–III

Magnetic Materials

1. On the basic of spin how the materials are classified as para, ferro, antiferro and ferri magnetic.

(i) Paramagnetic materials have few unpaired electron spins of equal magnitudes.

(ii) Ferro magnetic materials have manyunpaired electron spins with equal magnitudes.

(iii) Anti ferro magnetic materials have equal magnitude of spins but in anti parallel manner.

(iv) Ferrimagnetic materials have spins in anti parallel manner but with unequal magnitudes.

2. What is Bohr magneton?

The orbital magnetic moment and the spin magnetic moment of an electron in an atom can be expressed in terms of atomic unit of magnetic moment called Bohr magneton.

3. What is ferromagnetism?

Certain materials like iron(Fe), Cobalt(Co), Nickel(Ni) and certain alloys exhibit Spontaneous magnetization ie., they have a small amount of magnetization (atomic moments are aligned) even in the absence of an external magnetic field.This phenomenon is known as ferromagnetism.

4. What are ferromagnetic materials?

The materials which exhibit ferromagnetism are called as ferromagnetic materials.

5. What are the properties of ferromagnetic materials?

(i) All the dipoles are aligned parallel to each other due to the magnetic interaction between any two dipoles.

(ii) They have permanent dipole moment. They attract the magnetic field strongly.

(iii) They exhibit magnetisation even in the absence of magnetic field. This property of ferromagnetic materials is called as **spontaneousmagnetization.**

6. What is domain theory of ferromagnetism?

According to domain theory, a virgin specimen of ferromagnetic materials consists of a number of regions or domains which are spontaneously magnetized due to parallel alignment of all magnetic dipoles. The direction of spontaneous magnetisation varies from domain to domain.

7. **Mention the energies involved in origin of domains in ferromagnetic material.**

(i)   Magnetostatic energy

(ii) Crystalline energy

(iii) Domain wall energy

(iv) Magnetostriction energy

8. **What is antiferromagnetism?**

In anti-ferromagnetism, electron spin of neighbouring atoms are aligned antiparallel. Anti-ferromagnetic susceptibility is small and positive and it depends greatly on temperature.

9. **What are ferrites and mention its types.**

Ferrites are modified structure of iron with no carbon and in which the adjacent magnetic moments are of unequal magnitudes aligned in antiparallel direction. Its general formula is given by

$$X^{2+} Fe_2^{3+} O_4^{2-}.$$

Types: normally there are two types of structure.

(i)   Regular spinel

(ii) Inverse spinel.

10. **State the applications of ferrites.**

(i)   They are used in transformer cores for high frequencies uptomicrowaves.

(ii) They are used in ratio receivers to increase the sensitivity and selectivity of the receiver.

(iii) Ferrites are used in digital computers and data processing circuits as magnetic storage elements.

(iv) They are used as an isolator, gyrator and circulator which are used in microwave devices.

11. **What is hysteresis in magnetic materials?**

The lagging of magnetic induction (B) behind the applied field strength (H) is called hysteresis.

12. **What is meant by hysteresis loss?**

When the specimen is taken through a cycle of magnetization, there is a loss of energy in the form of heat. This is known as hysteresis loss

13. **What are soft-magnetic materials?**

Materials which are easy to magnetize and demagnetize are called soft magnetic materials.

14. State the properties of soft magnetic materials.

    (i) They have high permeability

    (ii) They have low coercive force.

    (iii) They have low hysteresis loss.

15. Mention few soft magnetic materials and their applications. Soft magnetic materials:

    (i) Pure or ingot iron

    (ii) Cast iron (carbon above 2.5%)

    (iii) Carbon steel

Applications:

    (i) Cast iron used in the structure of electrical machinery and frame work ofd.c.machine

    (ii) Carbon steel has high mechanical strength used in making motor of turbo alternators.

16. What ate hard magnetic materials?

Materials which retain their magnetism and are difficult to demagnetize are called hard magnetic materials.

17. State the properties of hard magnetic materials.

    (i) They possess high value of B-H product.

    (ii) They have high retentivity.

    (iii) They have high coercivity .

    (iv) They have low permeability.

18. What are ferromagnetic materials?

Materials which exhibit ferrimagnetism are called ferromagnetic materials. They are also known as ferrites.

19. Mention the properties of ferromagnetic materials.

    i) These are the ferromagnetic materials in which equal number of opposite spins with different magnitudes such that the orientation of neighbouring spins is in anti parallel manner.

    ii) Susceptibility is positive and very large for these materials.

20. Differentiate soft and hard magnetic materials.

S.No	Soft magnetic materials	Hard magnetic materials
1.	Magnetic materials which can be easily magnetized and demagnetized	Magnetic materials which cannot be easily magnetized and demagnetized
2.	They have high permeability	They have low permeability
3.	Magnetic energy stored is not high	Magnetic energy stored is high
4.	Low hysteresis losses due to small hysteresis	High hysteresis losses due to small hysteresis

21. Why ferrites are advantageous for use as transformer cores?

Ferrites are used as transformer cores for frequencies up to microwaves. This is because the eddy current problem which prevents the penetration of magnetic flux into the materials is much less severe in ferrites than in iron.

22. What is the origin of magnetic moment in magnetic materials?

The magnetic moment originates from the orbital motion and spinning motion of electrons.

23. What is diamagnetism?

When a material is placed in a magnetic field, the material becomes magnetized. The direction of the induced dipole moment is opposite to the externally applied magnetic field.

Due to this effect, the material gets very weakly repelled in the magnetic field. This phenomenon is known as diamagnetism.

24. What are diamagnetic materials?

The materials which exhibit diamagnetism are called diamagnetic material.

25. What are the properties of diamagnetic materials?

(i)   Diamagnetic materials repel the magnetic lines offorce.

(ii) There is no permanent dipole moment. Therefore, the magnetic effects are very small.

(iii) The magnetic susceptibility is negative and is independent of temperature and applied magnetic field strength.

26. What is paramagnetism?

In certain materials, net magnetic moment is zero though each atom or molecule possesses a permanent magnetic moment in the absence of an external magnetic field.

But when an external magnetic field is applied the magnetic dipoles tend to align themselves in the direction of the magnetic field and the material becomes magnetized. This effect is known as paramagnetism.

27. What are paramagnetic materials?

The magnetic materials which exhibit paramagnetism are called paramagnetic material.

28. What are properties of paramagnetic materials?

(i)   Paramagnetic materials attract the magnetic lines offorce.

(ii) They possess permanent dipole moment.

(iii) The susceptibility is positive

SUPER CONDUCTING MATERIALS

1. Define of super conductivity and super conductors.

The phenomenon of losing the resistivity absolutely to zero, when cooled to sufficiently low temperature ie., below critical temperature (T_C) is called superconductivity. The materials which exhibit superconductivity phenomena are called superconductors or superconducting materials.

2. What is transition temperature?

The temperature at which a normal material changes into a superconductor is called transition temperature (or) critical temperature (T_C).

3. Mention any four property charges that occur in super conductor (or) what are the properties of a superconductor.

(i) They have zero resistivity.

(ii) They exhibit perfect diamagnetism.

(iii) The super conducting property can be destroyed due to the application of magnetic and electric fields.

(iv) The transition temperature varies due to the presence of isotopes.

(v) The entropy and specific heat decreases at transition temperature.

(vi) The elastic properties, crystal structure and thermal expansion remains constant.

4. What is Meissner effect?

When a material is cooled below its transition temperature i.e., $T \leq T_C$, the material becomes a perfect diamagnetic. The magnetic flux originally present in the material gets ejected out of a superconductor. This effect is known as Meissner effect.

5. Explain the term critical magnetic field in superconductor.

At any temperature below the critical temperature, minimum magnetic field is required to destroy the superconducting property. This magnetic field is known as critical magnetic field (H_C). It is given by the relation, $H_c = H_o \left[1 - (T / T_c)^2\right]$

$H_C \rightarrow$ Critical magnetic field at any temperature $H_0 \rightarrow$ Critical magnetic field at absolute zero temperature $T_C \rightarrow$ Transition temperature of the material

6. What is isotope effect in superconductivity?

In a superconducting material, transition temperature varies with the average isotopic mass M of its constituents.

$$T_c \alpha \left[1 / M_\alpha\right]$$

Where α is called isotope effect coefficient.

7. **What are high TC superconductors? Give an example.**

Any superconductor, if transition temperature is above 10 K is called high TC superconductor. **Example:** $YBa_2Cu_3O_7$ $T_C = 92K$ $La_{1.85}Ba_{0.15}CuO_4$ $T_C = 36K$

8. **What are the properties of High TC superconductors?**

They have high transition temperature. They have modified pervoskite structure.

Formation of superconducting state in high TC superconductors is direction dependent. They are oxides of copper in combination with other elements.

9. **What are the applications of superconductors?**

Superconductors are used for the production of high magnetic field magnets. By using superconducting materials,it is possible to manufacture electrical generators and transformers in exceptionally small sizes having efficacy of 99.90% Superconducting materials are used in the construction of very sensitive electrical measuring instruments such as galvanometers. Superconducting materials if used for power cables will enable transmission of power over very long distances without any significant power loss or drop in voltage.

10. **What is magnetic levitation?**

The magnetic levitation is based on diamagnetic property of a superconductor which is the rejection of magnetic flux lines. A superconductor can be suspended in air against the repulsive force from a permanent magnet. This magnetic levitation effect can be used for high speed transportation without frictional loss.

11. **Distinguish between type – I and II superconductors.**

S.No	Type – I Superconductors	Type – II Superconductors
1.	The material loses magnetization suddenly.	The material loses magnetization gradually.
2.	They exhibit complete Meissner effect i.e., they are completely diamagnetic.	They do not exhibit complete Meissner effect.
3.	There is only one critical magnetic field (H_C).	There are two critical magnetic fields i.e., lower critical field (H_{C1}) and upper critical field (H_{C2}).
4.	No mixed state exists.	Mixed state is present.

12. **What is SQUID?**

SQUID is the acronym for Superconducting Quantum Interference Device. It is a double junction quantum interferometer. Two Josephson junctions mounted on a superconducting ring forms this interferometer. Squids are based on the flux quantization in a superconducting ring. The total magnetic flux passing through the ring is quantized. It is an ultra-sensitive measuring instrument used for detection of very weak magnetic field in the order of 10^{-14} tesla.

13. What is meant by persistent current?

When a d.c current of large magnitude is once induced in a super conducting ring, then due to the diamagnetic property of the super conductor, the magnetic flux is trapped inside the ring and hence the current persists in the ring for a longer time. This current is called as persistent current.

14. Define cooper pairs?

The pair of electrons formed due to the electron-lattice-electron interaction, with equal and opposite momentum and spins having the wave vector k-q and k'-q' are called Cooper pairs.

15. Define coherent length.

It is defined as the distance over which two electrons combine to form a cooper pair.

16. What is cryotron?

Cryotron is a type of switching element made by two different super conductors A and B as shown in fig(Refer cryotron fig in your book), with critical fields H_{cB} >H_{cA}. Here the super conducting property vanishes for the material 'A' due to the magnetic field produced by material B and hence it can be used as relay (or) switching elements.

17. Distinguish between A.C and D.C Josephson Effect.

S.N	D.C Josephson effect	A.C Josephson effect.
1.	When two super conducting materials are separated by an insulator of very few thickness and are connected by a wire, a d.c. current flows in the external circuit and is called d.c. Josephson effect.	When two super conducting materials are separated by an insulator of very few thickness and are connected to a d.c .power, thane an a.c. microwaves are produced at the junction and this effect is called a.c. Josephson effect.
2	The current persists for a longer time	The current persists only for a short time.

UNIT–IV

DIELECTRIC MATERIALS

1. Define dielectric constant?

It is the ratio between the absolute permittivity of the medium (ε) and the permittivity of free space (ε_0).

$$\text{Dielectric constant } \varepsilon_r = \frac{\text{Absolute permittivity } (\varepsilon)}{\text{Permittivity of free space } (\varepsilon_0)}$$

2. Define polarization of a dielectric material.

The process of the producing electrical dipoles inside the dielectric by the application an external electrical field is called polarization in dielectrics.

Induced dipole moment (μ) = αE E → Applied electrical field α → Polarizability

3. Name the four polarisation mechanisms.

    (i) Electronic polarisation.

    (ii) Ionic polarisation.

    (iii) Orientational polarisation.

    (iv) Space- charge polarisation.

4. What is electronic polarisation?

Electronic polarisation means production of electric dipoles by the applied electric field .It is due to shifting of charges in the material by the applied electric field.

5. What is ionic polarisation?

Ionic polarisation is due to the displacement of cations (negative ions) and anions (positive ions) in opposite direction due to the application of an electrical field. This occurs in an ionic solid.

6. What is orientation polarisation?

When an electrical field is applied on the dielectric medium with polar molecules, the dipole align themselves in the field direction and thereby increases electric dipole moment. Such a type of contribution to polarisation due to the orientation of permanent dipoles by the applied field is called orientation polarisation.

7. What is space- charge polarisation?

In some materials containing two or more phases, the application of an electrical field causes the accumulation of charges at the interfaces between the phases or at the electrodes. As result of this, polarisation is produced. This type of polarisation is known as space charge polarisation.

8. Define dielectric loss and loss tangent.

When a dielectric material is subjected to an A.C voltage, the electrical energy is absorbed by the material and is dissipated in the form of heat. This dissipation of energy is called dielectric loss.

In a perfect insulator, polarisation is complete during each cycle and there is no consumption of energy and the charging current leads the applied voltage by 90°.

But for commercial dielectric, this phase angle is less than 90^0 by an angle and is called dielectric loss angle. Tan is taken as measure of dielectric loss and is known as losstangent.

9. Define dielectric breakdown and dielectric strength.

Whenever the electrical field strength applied to a dielectric exceeds a critical value, very large current flows through it. The dielectric loses its insulating property and becomes conducting. This phenomenon is known as dielectric breakdown.

The electrical field strength at which dielectric breakdown occurs is known as dielectric strength.

10. Mention the various breakdown mechanisms.
   (i)   Intrinsic breakdown and avalanche breakdown
   (ii) Thermal breakdown
   (iii) Chemical and Electrochemical breakdown
   (iv) Discharge break down
   (v) Defect breakdown

11. What is intrinsic breakdown?

For a dielectric, the charge displacement increases with increasing electrical field strength. Beyond a critical value of electrical field strength, there is an electrical breakdown due to physical deterioration in the dielectric material.

12. What is thermal breakdown?

When an electrical field is applied to a dielectric material, some amount of hear is produced. This heat must be dissipated from the material. In some cases, the amount of hear produced is very large as compared to the heat dissipated. Due to excess of heat the temperature inside the dielectric increases and may produce local melting in the dielectric material.

During this process, a large amount of current flows through the material and causes their dielectric to breakdown. This type of breakdown is known as thermal breakdown.

13. What is chemical and electrochemical breakdown?

Electro chemical breakdown is similar to thermal breakdown. When the temperature of a dielectric material increases, mobility of ions increases and hence the electrochemical reaction may take place. This leads to leakage current and energy loss in the material and finally dielectric breakdown occurs.

14. What is discharge break down?

Discharge breakdown occurs when a dielectric contains occluded gas bubbles. When this type of dielectric is subjected to electric field; the gases present in the material will easily ionize and thus produces large ionization current. The gaseous ions bombard the solid dielectric. This causes electrical deterioration and leads to dielectric breakdown.

15. What is defect breakdown?

The surface of the dielectric material may have defects such as cracks, porosity and blowholes. Impurities like dust or moisture may collect at these discontinuities (defects). This will lead to a breakdown in a dielectric material.

16. What are requirements of good insulating materials?

The good insulating materials should have

(i)   High electrical resistivity to reduce leakage current.

(ii) High dielectrical strength to with stand higher voltage.

(iii) Smaller dielectric loss

(iv) Sufficient mechanical strength.

17. Compare active and passive dielectrics.

S.No	Active dielectrics	Passive dielectrics.
1.	Dielectrics which can easily adapt itself to store the electrical energy in it is called active dielectrics.	Dielectric which restricts the flow of electrical energy in it are called passive dielectrics.
2.	Examples. Piezo electric ,Ferro electrics	Examples. Glass, mica, plastic
3.	It is used in the production of ultrasonics.	It is used in the production of sheets, pipes, etc.,

18. What are ferro-electric materials? Give examples.

Materials which exhibit electronic polarization even in the absence of the applied electrical field are known as ferro-electric materials.

Example.

Barium Titanate ($BaTiO_3$)

Potassium Dihydrogen Phosphate (KH_2PO_4)

19. What are the differences between polar and non-polar molecules?

S.No	Polar molecule	Non-polar molecules
1.	These molecules have permanent dipole moments even in the absence of an applied field.	These molecules do not have permanent dipole moments.
2.	The polarization of polar molecules is highly temperature dependent.	The polarization of polar molecules is temperature independent.
3.	These molecules do not have symmetrical structure and they do not have centre of symmetry.	These molecules have symmetrical structure and they symmetry.
4.	For this kind of molecules, there is absorption or emission in the infrared range.	For these molecules, there is no absorption or emission in the infrared range.
5.	Examples: $CHCl_3$, HCl	Examples: CCl_4, CO_2

20. What is meant by pyro-electricity?

It means that, the creation of electronic polarization by thermal stress.

Unit-V

Modern Engineering Materials

1. What are metallic glasses?

Metallic glasses are the newly developed engineering materials which shares the properties of both metals and glasses. They are glasses having metallic properties.

2. What are the types of metallic glasses and mention few metallic glasses.

There are two types of metallic glasses, they are:

(i) Metal –Metalloid metallic glasses

(ii) Metal –Metal metallic glasses

Metal–Metalloid metallic glasses (Fe, Co, Ni) Metal – (B, Si, C, P) Metalloid

Metal–Metal metallic glasses Nickel–niobium (Ni – Nb)

3. State the structural properties of metallic glasses.

(i) They do not have any crystal defects such as grain boundaries dislocationetc.

(ii) Metallic glasses have tetrahedral close packing (TCP) in contrast to hexagonal close packing (HCP) of the crystalline solid.

4. What are the mechanical properties of metallic glasses?

(i) Extremely high strength due the absence of point defects, dislocation and slip plane.

(ii) They have high elasticity.

(iii) They are highly ductile

5. What are the electrical properties of metallic glasses?

(i) Electrical resistively of metallic glasses is high (> $100\mu\Omega$ cm) and it does not vary much with temperature.

(ii) Due to high resistivity, the eddy current loss is verysmall.

6. What are the magnetic properties of metallic glasses?

(i) Metallic glasses have both soft magnetic and hard magneticproperties.

(ii) They exhibit high saturation magnetization.

(iii) The core losses of metallic glasses are veryless.

7. What are the chemical properties of metallic glasses?

(i) They are highly resistant to corrosion due to the formation of protective oxide film in chromium containingglasses.

(ii) They are highly reactive and stable.

8. **What are the applications of metallic glasses?**

   (i) Metallic glasses possess high tensile strength. They are superior than common steels. This makes them useful as reinforcing elements in concrete, plastic or rubber.

   (ii) Due to their high strength, high ductility, rollability and good corrosion resistance, they are used to make razor blades. This fact is also utilized to make different kinds of springs.

9. **What are the advantages of using metallic glasses as transformer core material?**

Metallic glasses are ferromagnetic. They possess low magnetic losses, high permeability and saturation magnetization with low coactivity. They also have extreme mechanical hardness and excellent initial permeability. These properties make them useful as transformer core materials. Moreover power transformers made of metallic glasses are smaller in size and efficient in their performance.

10. **What are shape memory alloys?**

The group of metallic alloys which demonstrate the ability to return to their original shape or size (i.e., the alloy appears to have memory) when subjected to the appropriate thermal procedure (heating/cooling) is called Shape Memory Alloys (SMAs).

11. **What is shape memory effect?**

Certain metallic alloys like alloy of gold (Au) and Cadmium (Cd) exhibit a plastic nature, when cooled to a lower temperature. The return to their original dimensional configuration (metallic) during heating at high temperature. This effect is called Shape Memory Effect (SME).

12. **What are the properties of shape memory alloys?**

   (i) They can exist in two different solid phases with distinct crystal structures in SMA.

   (ii) If temperature is increased, material goes to austenite phase which has cubic crystal structure and on cooling; the material comes back to its original shape in the martensite phase.

   (iii) SMAs exhibit changes in electrical resistance, volume and length during the transformation with temperature.

   (iv) They are extremely elastic or exhibit **pseudo elasticity,** i.e., strain can be very large for a given stress in the martensite phase.

   (v) SMAs exhibit self-healing effect.

13. **What are the applications of shape memory alloys?**

   (i) Shape memory alloys can act as actuators and sensors.

   (ii) Fiber composite shape memory alloys are used to produce twist on the helicopter blades.

(iii) They are used in orthopaedic devices for pulling fractures together, artificial hearts and shrink-wrap.

14. What is glass transition temperature?

The temperature at which the transition from liquid (metallic liquid) to solid (glass) occurs is known as the glass transition temperature.

15. What are nano phase materials?

Nanophase materials or nanaomaterials are newly developed materials with grain size at the nanometer range (10-9), i.e., in the order of 1–100 nm.

The particle size in a nano materials is 1 nm.

16. Mention different forms of nanomaterials.

Nanodots, nanorods, Carbon nanotubes and Fullerenes.

17. What are two routes through which nano particles can be synthesized?

(i) **Top down approach:** involving breaking down bulk materials tonanosizes.
   Example: Mechanical alloying
(ii) **Bottom up approach:** where the nano particles are made by building atom by atom.
   Example: Inert gas condensation

18. Mention few techniques for synthesis of nano phase materials.

(i) Mechanical alloying
(ii) Inert gas condensation
(iii) Sol-gel technique
(iv) Electro-depostion
(v) Laser synthesis
(vi) Spraying

19. What are physical properties of nanomaterials?

Melting point reduces with decrease in cluster size.

Interparticle spacing decreases with decreases in grain size for metal clusters.

Ionization potential changes with cluster size of the nano grains.

Greater luminescence efficiency in nano semiconductor materials.

20. What are mechanical properties of nanomaterials?

Higher hardness and mechanical strength (2-7 times) when grain size reduces from 1 μm to 10 nm. Higher moduli of elasticity (30%-40%), Very high ductility and super plastic behavior at low temperatures.

21. What are magnetic properties of nanomaterials?

Non-magnetic materials become magnetic when the cluster size reduces to 80 atoms.

Bulk magnetic moment increases with decreases in coordination number Ferro magnetic materials exhibit super paramagnetism at nanograin sizes. Paramagnetic materials exhibit ferromagnetism at nano grain size.

22. What is non-linear optics?

The field of optics dealing with the non-linear behavior of optical materials.

23. Name few non-linear optical phenomena.

The few of the nonlinear phenomena observed are:

(i)   Second harmonic generation

(ii) Optical mixing

(iii) Optical phase conjugation

(iv) Soliton

PART–B QUESTIONS

UNIT-I

CONDUCTING MATERIALS

1. Deduce a mathematical expression for electrical conductivity and thermal conductivity of a conducting metal and hence, obtain Widemann-Franz law.

2. Define Fermi energy, and derive an expression for the Fermi energy of a system of free electrons.

3. i) Define density of states in metals in

   ii) Write down the expression for Fermi-Dirac distribution function.

   iii) Derive an expression for the Fermi energy of a system of free electrons.

4. With a neat diagram and derive an expression for density of states.

5. Write Fermi-Dirac distribution function. Explain how Fermi function varies with temperature.

6. With the help of Fermi-Dirac statistics, derive the expression for density of states and deduce Fermi energy.

7. i) What are the special features of classical free electron theory?

   ii) Derive an expression for the electrical conductivity of a metal.

   iii) How it is affected by temperature and alloying?

8. Derive an expression for density of states in a metal and hence obtain the Fermi energy in terms of density of free electrons, at 0K.

UNIT–II

SEMICONDUCTING MATERIALS

1. Assuming Fermi-Dirac statistics, derive expressions for the density of electrons and holes in an intrinsic semiconductor. Hence, obtain expression for the electrical conductivity of the intrinsic semiconductor.

2. Derive an expression for,
    i. Density of electrons in the conduction band in an intrinsic semiconductor.
    ii. Density of holes in the valance band in an intrinsic semiconductor.

3. Describe the method of determining the band gap of a semiconductor. How does the electrical conductivity vary with temperature for an intrinsic semiconductor?

4. Obtain an expression for, density of electrons in the conduction band of an N-type semiconductor and density of holes in the conduction band of a P-type semiconductor by assuming Fermi-Dirac distribution function.

5. Discuss the variation of carrier concentration with temperature in n-type semiconductor.

6. Derive an expression in intrinsic semiconductor.

7. i) Derive an expression for the carrier concentration in intrinsic semiconductor.
    ii) Explain the variation of electrical conductivity with respect to temperature in the case of an intrinsic semiconductor. Describe a method of determining the band gap energy of a semiconductor.

8. Derive an expression for carrier concentration with temperature N-type semiconductor, and describe an experimental set-up for the determination of Hall Co-efficient and Hall voltage.

9. Explain the variation of Fermi level wit

10. Explain the variation of Fermi level with temperature and impurity concentration in N-type semiconductor.

11. What is Hall effect? Describe an experiment for the measurement of Hall coefficient, and write its applications. Derive an expression for density of holes in the valance band and also explain how does the Fermi level vary with concentration of impurities in P-type semiconductor.

UNIT-III

MAGNETIC MATERIALS AND SUPERCONDUCTING MATERIALS

1. Write a note on magnetic storage.

2. Explain in detail about the magnetic memories.

3. i) What is ferromagnetism?

   ii) Explain the reason for the formation ofdomain structure in ferromagnetic material and how the hysteresis curve is explained on the basis of domain theory?

4. i) Classify the magnetic materials on the basis of their spin?

   ii) What are ferromagnetic domains? Explain.

5. Explain the domain theory of ferromagnetism. Using that how will you explain the properties of ferromagnetic materials?

6. i) What are ferromagnetic domains?

   ii) Draw a B-H curve for a ferromagnetic material and identify retentiveand coercive fields on thecurve.

   iii) What is energyloss per cycle?

   iv) What are ferrites?

7. What are hard and soft magnetic materials? Compare their properties. Give some examples.

8. Distinguish between dia, para, ferro and ferromagnetic material. Mention their properties and applications.

9. Explain the different structure of ferrites with suitable diagrams.

10. Write a short note on the following

    i) Anisotropyenergy

    ii) Magnetostrictive energy

    iii) Domainwallenergy

    iv) Exchange energy.

11. i) Explain ferromagnetism based on domain theory.

    ii)What are the four types energy involved in domain growth? Explain.

12. Discuss the domain structure in ferromagnetic materials.

13. Discuss the properties and applications of superconductors?

14. i) Distinguish between Type –I and Type – II Superconductors.

ii) Describe hightemperature superconductors.

15. i) What is super conductivity?

ii) Mention any four properties of superconductors.

iii) Explain the effect of isotopesin superconductors.

iv) Explain type-I and type –IIsuperconductors.

v) Mention tow applications.

16. i) Explain Meissner effect and magnetic levitation.

ii) Discuss the applications of superconductors.

17. Mention the applications of superconductors.

18. Discuss type-I and type-II superconductors.

19. Explain superconducting phenomena. What are its properties? What are its Applications?

20. Distinguish type-I and type-II superconductors Write a note on high temperature superconductors.

21. Write an essay on superconducting materials and its applications. What are the new developments?

22. Explain the characteristics of superconducting materials.

23. Write a short note on

24. i) Meissner effect   ii) Josephson effect

iii) Magnetic levitation   iv) SQUID

25. Based on tunneling explain the A.C. and D.C. Josephson effects and its applications in the superconductors.

Unit-IV

Dielectric Materials

1. What is meant by polarization in dielectrics? Derive the relation between the dielectric constant and atomic polarisability.

2. Explain the different types of polarization mechanisms in a dielectric.

3. Explain the frequency and temperature dependence of various polarizations.

4. What is meant by local field? Derive an expression for Lorentz field in a dielectric material and hence deduct Clausius - Mosotti relation.

5. What is dielectric loss. Derive an expression for power loss in a dielectric.

6. Discuss in detail the various dielectric breakdown mechanisms.

Unit-V

Advanced Engineering Materials

1. Give a detailed account of metallic glasses, their method of production, types, properties and applications.

2. What are shape memory alloys? Write their characteristics. List out the applications of shape memory alloys. What are its advantages and disadvantages?

3. Describe with neat sketch the plasma arcing processes to produce nano materials system.

4. Explain about chemical vapour deposition technique for the fabrication of Nanoparticles.

5. Explain in detail the Metal Organic Vapour Phase Epitaxy [MOVPE] technique for the fabrication of nanoparticles

6. Describe liquid phase chemical synthesis to produce nano particles

7. Describe the colloidal and sol-gel methods for producing nano crystals

8. Describe electro deposition process to produce nanoparticles.

9. Describe a ball milling technique with neat diagram to produce nanomaterials.

10. Explain the various properties of nanomaterials.

11. Describe the application of nanomaterials.

12. What are carbon nanotubes? Explain the types, fabrication, properties and applications of carbon nanotubes.